AUSTRALIAN WOOLSHEDS

Woolshed interior by Luke Sowden, aged seven

AUSTRALIAN WOOLSHEDS

Edited & photographed by Harry Sowden

CASSELL AUSTRALIA

CONTENTS

The Editor would like to thank the following for their help in the preparation of this book;
Wayne McPhee for researching the history and development of the Australian wool industry;
Georgina Blyth for assisting him;
Margaret Sowden for co-ordinating the material;
and all owners and managers of properties who kindly helped with information and hospitality.

CASSELL AUSTRALIA LTD
30-36 Curzon Street, North Melbourne, Victoria
80 Bay Street, Broadway, New South Wales

First published October 1972
Reprinted April 1973
ISBN 304 93922 6
Registered in Australia for transmission by post as a book
Designed by Harry Sowden
Printed and bound by John Sands Pty. Ltd.
473

JELLALABAD POLWARTHS

INTRODUCTION

If Australia can boast of a vernacular architecture, then it must surely be in its woolsheds. For these buildings have been shaped by Australian history. They grew from the humbleness of the pioneer to the grandeur of the wealthy grazier, a chequered development that fluctuated with the economy.

Today, with the uncertainty of wool prices, the economy again plays its part and the fate of the woolshed is in the balance. Investment is fickle as the country greedily devours itself in search of a new wealth. In a hard country with little time for sentiment, the large woolsheds of yesteryear, that once resounded with the activity of up to two hundred men at shearing time, now fall into disrepair. Many others have been destroyed by time or fire.

This book sets out to record this unique era of Australian architecture before its page of history is closed for ever. To cover it fully would take a lifetime, as there are hundreds of woolsheds throughout Australia, scattered over thousands of square miles. Those chosen aim to show the variety of the woolshed, in materials, size and character. Here is a range of workaday buildings of straightforward functional design that illustrate an important, though largely ignored, aspect of the architecture of their time.

Harry Sowden
Sydney, 1972

Out of this grew the indigenous Australian woolshed

History has shown man's strong preoccupation with the grazing of sheep for food and clothing, to the extent that he owes much to these animals (and especially to their fleeces) for his mastery over the elements and the attainment of civilization. The long association sheep flocks have had with man has today made sheep development more a product of human engineering than of natural evolution.

The early history and development of the woollen industry is so well hidden in the past that it is not known when sheep were first domesticated and grazed, although Neolithic man possessed them, and woollen clothing and sheep are closely related to the earliest trade alliances. Of the ancient civilizations, it is known that the Egyptians grazed very large sheep flocks from which they obtained fine woven material, as did the Phoenicians. It is due to this latter race of people that large quantities of woollen materials and sheep were traded and thus introduced into many European countries including France and, importantly, Spain.

Little is known of the methods of shearing and tending these early flocks, or of the implements and architectural forms that were to result from the necessities of shearing; however, archaeological research has fortunately provided us with the remains of various shearing tools. From less sophisticated methods of plucking the wool from the sheep by hand, or cutting the wool with an edged sickle, can be seen the development of wide and thin bladed scissors (with blades riveted together) and bow shears rather like garden clippers. Various types of the latter were developed in the last century for use by Australian shearers. Often sheep were shorn in the open on a cloth laid down to collect the wool, the task being performed by a number of people.

With larger flocks, and in countries where protection from the weather was essential, buildings were erected for the purpose of shearing. These buildings, which were rudely constructed from local materials, also served as storage for grain and fodder.

The earliest written records of sheep and shearing sheds may be found in the Bible, but mention is unfortunately made only indirectly.

> Genesis XXXI:19: '. . . Laban went to shear his sheep' and II Samuel XIII:23, 28: '. . . and it came to pass that Absalom had sheepshearers in Baal-Hazor . . . and Absalom invited all the King's sons. . . . Now Absalom had commanded his servants, saying, "Mark ye now when Ammon's heart is merry with wine . . . then kill him".'

There was great merriment and celebration (as well as evil plotting) associated with the sheep industry of the Old Testament times and the shearing shed was very much a part of it.

> II Kings: 13, 14: Jehu said, '"Take them [the brethren of Ahaziah] alive." And they took them alive, and slew them at the pit of the shearing house. . . .'

i.e. the pit used for washing sheep before shearing took place.

It was not until well after the grazing of sheep had become a national characteristic of the Spanish that woolsheds are referred to specifically and in detail. The following description from *Travels Through France and Spain*, by Maurice Keatinge, in 1817, gives one some idea of the scale of both the wool industry and shearing sheds in Spain, which by this time were highly developed.

> These buildings usually enclose large courts of which some are of the dimensions of six hundred feet by four hundred; with every accommodation in the upper storey for lodging the master, his family and the persons requisite for the attendance on the flock at shearing time, which takes about a month; the rest of the year these places are left in care of a servant and otherwise are unoccupied.[1]

Shearing time was a highly organized affair. Large

flocks travelled overland to big centralized sheds to be shorn and the work-force employed was enormous, some 40,000 persons to attend 5,000,000 sheep. The travelling flocks were known as *Transhumantes*, the stationary ones the *Estantes*. The former were more valuable and made an annual 500-mile migration to be shorn. Shearing was supervized by a master and shearers were aided by women pickers who gathered up the loose wool, basket carriers who collected it, *moreneros* (boys with vessels of charcoal to rub the galled sheep) and *echavinos* (wine servers) who went around with drink for all, each person being allowed eighteen draughts per day. Workers, divided into squads, were supervized by chiefs so as to maintain order in the work, which was carried out in a very industrious yet festive manner. Each man's daily ration consisted of mutton (one sheep being allotted to every ten men) and two pounds of bread. Work began at six in the morning and finished at five in the evening. The shorn fleeces were first weighed then sorted into three different grades on timber frames. Cleaning then began. This consisted of seven separate steps: washing, draining, collecting, washing again, treading, draining and finally drying in the sun. It is no wonder that, with as rich and highly organized an industry as this, Spain guarded her flocks jealously. However, sole control of the industry was lost with French invasion and Spanish merinos were shipped abroad to Saxony and Silesia. By the turn of the nineteenth century many countries were importing large quantities of wool from France and Germany.

Great Britain had unsuccessfully tried to emulate the Spanish wool growers as early as the fifteenth century and develop an industry of her own. However, despite this early lack of success, George III acquired several rams and 36 ewes as a gift from the Marchioness del Campe di Alange between 1787 and 1791. This caused so great an interest in England at the time that by 1811 a Merino Society had been properly formed, in order to encourage the breeding of these sheep. All aspects of the Spanish methods were carefully studied, from grazing to shearing, so as to ensure success. A well-known botanist, whose association with early Australian history is linked in almost equal terms with that of Captain Cook, was the Society's president, Sir Joseph Banks. Banks was an enthusiast for merino breeding and the only man of the time to command a detailed knowledge of the Australian environment, though, somewhat surprizingly, he was at first unsure of the merinos' suitability for the Australian climate. It was eventually left to the resource of Captain Macarthur to establish the merino in Australia and begin a new era in world sheep history.

In a time when convicts and officials alike were struggling with the enormous difficulties of carving an existence in a wide continent, and dependent on a precarious supply of goods from England, the first merino fine-woolled sheep were imported. In 1796, when ships were dispatched to the Cape of Good Hope for supplies, Macarthur asked Captain Waterhouse and Captain Kent to purchase any fine-woolled sheep that opportunity offered. In this venture they were successful, as the Spanish king had some time previously presented the Dutch Government with pure-bred merino sheep under the care of Colonel Gordon, which were sold after Gordon's death. Kent and Waterhouse purchased 26 merinos and shipped them back to Sydney, little realizing the full importance of their precious cargo. Sheep had previously been brought to the colony for their meat value. According to early records these were fat-tailed sheep collected at Cape Town. The fat, broad tail was the favourite part of the sheep and the fat was used in place of butter. Later, sheep were imported from Bengal (a poor class of sheep in comparison with the merino) but still bred for mutton. The remaining few not used for this purpose were more selectively bred until the strain, in time, was greatly improved.

On the arrival of Kent and Waterhouse in 1797, Captain Macarthur immediately saw the value in the few sheep that had survived the arduous sea journey, and offered £15 per head for them. This offer was not immediately accepted, but Macarthur eventually obtained three rams and five ewes and crossed them with his own sheep. This, Australia's first wool flock of merino strain, was grazed on Macarthur's property, Elizabeth Farm, near Parramatta. No records are available to describe the shearing methods, but the first wool was removed from these sheep on the grassy fields in front of the homestead, in the open. In 1800 eight sample fleeces were sent to England for inspection and report by Sir Joseph Banks. The first proof of Australia's future success in the wool industry came in 1808 when a bale of 245 lb. of wool, grown by Macarthur, was sold in London after highly competitive bidding. By this time, Macarthur had been granted 5,000 acres of grazing land and had obtained from George III's stud at Kew seven rams and one ewe of the Negretti strain. He had set in motion his vision of an Australian pastoral industry with a measure of self-sufficiency for the colony.

Another man full of enthusiasm for the wool industry was the Reverend Samuel Marsden. Marsden also obtained several merinos imported by Captain Waterhouse and grazed his sheep on his farm at Botany Bay. Marsden crossbred his sheep and experimented in developing better breeds. Again, few records are available to give a full picture of this, the beginning of the wool industry, although it is known that early consignments of wool (which reached England in 1808) were sent in wooden barrels pressed in tightly with a spade.

Early statistics show that sheep numbers in the colony had increased from 29 in 1788 to 6,124 by 1800, so great was the interest taken in sheep-farming. The early success of wool crops grown by Macarthur and Marsden inspired many free settlers, the first of many land grants being made in 1792 to the emancipated convict, James Ruse. Later, in 1830, this system of land division was ceased and auctioning was established by the Government.

On thickly forested coastal stretches there grew the first rough homesteads and outbuildings. These were constructed by means of timber posts and beams, with frames of smaller members to enable the fixing of bark walls and roofing sheets. Local timber was cut and logs trimmed to form posts, beams, roofing ties and riders, the riders being secured over the roof to prevent wind from lifting the bark sheets.

Farming land was at first confined generally to the coastal areas around the colony where native trees were felled to make way for pastures, the timber being used for the homestead and outbuildings and the surplus usually burnt. Timber was the most abundant material and in most cases easily felled and split. The first farm buildings were completely of timber construction, with cane grass, blackboy rushes or bark for the roof. A straw and clay mixture was usually plastered between cracks of the slab timber walls.

Successful exploration west of the colony did not occur until 1818, when Wentworth, Blaxland and Lawson crossed the Blue Mountains to open up vast pasture lands. Before the opening up of the inland and coastal areas, sheep-farming took place on the heavily wooded areas adjacent to the settlement and was of a small scale by comparison with the later land boom.

The first settlers brought with them sheep-farming methods developed in the Old Country which were, in turn, a slight variation on the Spanish farming methods mentioned previously. Often a large portion of a settler's luggage contained his farming implements as well as books on farm management and home remedies, as farm implements were in short supply in the colony and ships visited all too infrequently.

Flocks at first were small and convicts were employed on low 'wages' as shepherds and farm-hands. A shepherd cared for the sheep during the day and herded the animals into enclosures of timber hurdles (made from split palings) during the night. As flocks grew, and large sheep runs became established in what was to become Queensland and South Australia, many shepherds were needed to protect enormous flocks. Portable, demountable hurdles were built from light timber for this purpose and were so designed as to be easily reassembled. Moved in by dray, they were quickly assembled around the flock in the evening. It was common for the shepherd to sleep near the gate of the yard in a watch box. This was an enclosure made from bark and scantling poles. About seven feet long by four feet wide, it contained two openings, one in which to enter the kennel-like structure on poles and the other smaller, yet sufficiently large to enable the shepherd to view the sheep clearly. The shepherd's lonely lot was in later years made more comfortable by the construction of larger, more permanent dwellings.

At shearing time convicts were employed, usually for woolshed jobs (washers, shearers, balers and shed rouseabouts) under the watchful eyes of the property owner, who would take part in these operations as well as supervize. Many convicts were chosen by property owners because it was known that they had been employed in similar farming pursuits in England and, although shearing method and shed design were improved and developed in different ways, initially tradition prevailed. The increase in flock sizes, improved notions on animal behaviour and particular environmental conditions, led to a wide variety of shed design that expressed all these factors in superb economy and practical use of local materials. The various traditional shearing methods practised in the colony in early days, however, were just as interesting and diverse.

Many property owners followed the English method of cleaning the fleeces prior to shearing. This simply consisted of compelling the sheep to swim across a running stream a number of times until the wool was observed to be clean and free from grease. In many European countries the Spanish method was followed, of washing and cleaning fleeces after shearing. The wool was immersed in large pits of hot water, the various layers trodden on, then taken out and drained on 9 ft. x 4 ft. 6 in. flat timber frames. This process was repeated in cold running stream water then, after draining on the frames and the wool being squeezed by hand, the fleeces were dried in the open fields. Various farmers, because of the absence of streams, chose to construct wooden pens in the nearest large local waterhole or billabong. These pens were built of floating logs lashed to timber piles driven into the bed of the waterhole. A sheep race or narrow corridor directed the sheep along from catching pens and down a slippery-dip ramp into the water pen. It was then a matter of ducking the sheep under the floating logs into a similar second pen where the sheep washers squeezed the grease and dirt out by hand, in turn ducking the sheep into yet another water pen, this one being narrower and up which the sheep swam to solid ground and to yards on the opposite bank. The sheep were usually both alive and dry at the conclusion of these operations.

On most properties it was for the shearers to perform the task of washing. The first three days of the week were usually spent washing and cleaning and, on the fourth, the first sheep washed (the driest) were shorn. This process was continued till all the sheep were shorn and the wash pool empty. Shearing was usually performed any time from May to December and which particular month suited the farmer depended on many factors such as drought, the advent of flood, or other farm activities.

Early sheds were simple wooden structures consisting of one large area, rectangular in plan, with trodden earth or dried leaves for a floor. The roof was pitched in both directions; later additions were at a slightly lower angle to the main hipped roof to provide both water run-off and head-room at the eaves. Often end walls and one of the long walls would be left free of cladding so that sheep and stores could be more easily moved to and from the enclosure. Shed hands

soon learned how to construct intricate pens, races and gate systems to encourage sheep to move easily into the shed work areas and, after shearing, into the open fields.

The main activity was centred on the board, an area in the centre of the shed where the shearer removed the fleece. On either or both sides of the board were located catching pens, from which the shearer selected his sheep. Those with soft wool and bare bellies, easily shorn, were usually the first, leaving the more difficult until last. Many plan variations were developed from this principle of selecting a sheep, shearing the animal and sending the fleece and sheep on their respective ways, and will be referred to in detail later.

Each shearer's work space contained rough furniture, usually a plank or length of timber nailed to a nearby post, on which to store items essential to the job such as a sharpening stone, an extra pair of shears, his tin of tobacco, pipe, matches and tally marker. It became a custom very early in Australian sheep history to pay the shearers for the number of sheep shorn, rather than on an hourly basis, for obvious reasons mainly advantageous to the property owner. A strong sense of competition, by which Australians are universally characterized, soon developed on the shearing board. Many ingenious methods were devized to record the number of sheep each shearer had shorn for the day, ranging from notches on the edge of the shelf containing his shearing tools to dropping a marble into an empty jam tin for each ten sheep shorn. Each man would mark his sheep after it had been shorn with a piece of red raddle (or ochre) on the tail or thigh, so as to distinguish his tally from the other shearers' in case of dispute and to enable the boss to find culprits who had cut sheep badly. The latter were marked with blue raddle by the property owner and an appropriate amount deducted from the shearer's wages. It goes without saying that the board overseer would at all times have to ensure that these tallies were properly recorded.

The methods of shearing varied. Many stuck to

tradition and employed the labour of several people to each sheep, two to hold the animal while another shore or clipped the wool off on to a sheet laid down for that purpose, and others to gather the loose wool and store it in separate lots before placing in barrels or cloth bales slung from timber beams. Many shearers used a curved knife in the manner used in North Africa, while tales related tell of ex-tailors from London shearing sheep. In times when labour was scarce, in the late 1830s in South Australia, many young German women immigrants would shear the sheep.

> . . . the young woman—without shoes or stockings—had a short piece of thick string tied to her great toe, the other end being tied to the hind feet of the sheep. The girl's leg was then stretched to extend to the legs of the sheep. Her knee or left hand was placed on the neck or shoulder of the animal, which was then left to her charge and she commenced her clipping, most carefully avoiding any snips of the skin. The number shorn was not more than thirty a day. The wool was not taken off very close.[2]

Up to the 1830s there was very little organization during shearing time. Shearers (those additional to the farm's convict labour) heard of work by word of mouth, usually at the local hotel. They were issued with shears, and tar ointment for cuts inflicted on sheep, after arrival at the shed on the arranged day of starting. Work hours stretched from daybreak to sunset with lunch and frequent smokoes in between, payment being measured in the number of scores (not hundreds as in later years) of sheep shorn. A good tally was 80 sheep per day and the shearer who shore 90 sheep or above would be the shed's top hand, or ringer. Shearers of this era did not know the luxuries of shearers' cooks, tar-boys, rouseabouts or shearers' huts. These were to follow after the exploration into western New South Wales and further reaches of the Australian continent.

By the time the last remaining sheep, or cobbler, had been shorn and the wool packed in bales and loaded on to horse or bullock-drawn drays for the city, the shearers would have received their pay (those who were not convict labourers) and removed to the nearest hotel in order to spend as many of their well-earned pounds as possible. They then headed off in search of more work. The property owner could relax in the knowledge that twelve months' work had come to fruition and the silent shed was swept clean to be returned to its other farm purposes—storage, shelter, workshop. All energy was then directed towards making successful the next year's wool harvest.

The merino's suitability for Australian climatic conditions soon led to enormous increases in the size of stock runs. Whereas previously the graziers would travel with their flocks, tend, shear and send the wool away by bullock drays, now convicts and ticket-of-leave men were employed to do many of these jobs, and the organizational changes resulted in a different scale of operation and a revolution in industrial methods. Increase in flock sizes during the 1830s coincided with great interest being taken in England. By this time bankers were convinced that the colony's growing industrial empire was a suitable place to invest funds, despite the largely convict population and the rugged breadth of the land. There was also rapid increase in the flow of immigrants, mainly due to economic depression felt in some parts of England. These factors, combined with rapid exploration and opening up of large tracts of land to the north, west and south of Sydney, led to a land rush which, between 1838 and 1847, was to be barely surpassed by the gold rush of the next decade. Free settlers from England were aided by the Government in establishing their farming land, and could select large areas of virgin bush where they wished. Other less financially endowed settlers, ticket-of-leave men and pioneers simply followed closely behind explorers or became amateur explorers. These pioneers squatted on the largest and best grazing land available to them and earned contempt from the free selectors.

Soon after the Blue Mountains had been traversed, exploration teams began moving into the northern and southern reaches of the country. In 1824, explorers Hamilton Hume and William Hovell travelled south from Sydney and crossed the Murray River to open up some of the richest country in the Australian continent. Their initial assessment of the land was that it would be 'highly suitable for sheep grazing'. This later proved accurate and, judging by results in later years, almost an understatement. The first white settler to begin sheep farming in Victoria, Edward Henty, met with enormous success. Explorer Major Mitchell told Henty in 1836 of very rich pastures on the banks of the Wannon River and Henty settled this, calling it Merino Downs.

It was not long before news of these rich southern pastures spread. Large numbers of graziers from New South Wales moved their flocks south across the Murray River to establish their runs, which were mostly in the Western District of Victoria. In 1835, a pioneer and colourful figure of this time, Mrs W. Furlonge, transferred her sheep to Seven Creeks, Euroa, Victoria, where she established a merino stud of Saxony blood which greatly helped to boost the wool industry in the south. Between 1836 and 1840 sheep numbers in Victoria had risen from 41,332 to 782,283. The fervent pioneering activity and expansion of the 1830s and 1840s can be gauged by comparing the 782,283 sheep in Victoria in 1840 with the 3,000,000 that were grazed there in 1847. By 1851 this figure had almost doubled.

Prior to this, however, in 1841, a slump occurred in the wool trade which caused a depression in the pastoral industry. Melbourne was the most severely affected place and many sheep farmers were forced to press further into the interior of the country towards the South Australian border, boosting the development of the industry in South Australia which was begun by the South Australian Company in 1836. Saxony merinos had been first introduced into South Australia in 1837 and, by 1844, the South Australian Company owned 31,848 sheep as well as a large number of pure-bred merinos. Farmers soon learned which breed of sheep best suited the climate in each locality—pure-bred merinos in Queensland, where there was a relatively higher rainfall, and a more robust merino in the drier regions of South Australia. Fleeces won from South Australian sheep were called red wools, due to the red sand swept by wind over vast areas of grazing land, and the flat, dry, dusty environment of South Australia also forced a different approach to farm, homestead and outbuilding design.

In the early years Queensland formed part of New South Wales. It was known as Moreton Bay and its sheep-farming development at first closely followed that of New South Wales. Fine-woolled sheep were first introduced in 1823 and were owned by the Government, Rambouillet merinos being introduced in 1825. In the north of New South Wales, large farming lands were opened up on the Darling Downs. In 1840 Patrick Leslie, a young grazier from the Upper Hunter following in the tracks of explorer and botanist Allan Cunningham, came across excellent grazing land at the headwaters of the Condamine River, where he made a head station near where the town of Warwick now stands. By 1842, the country had been opened for private settlement and flocks could be acquired from the Government. In this same year, Governor Sir George Gipps visited Moreton Bay, where a scattering of 45 stations had been established, and twelve months' produce of 1,800 bales of wool had been shipped to Brisbane.

The coastal climate of Queensland proved unsuitable for the merinos. In later years graziers moved their flocks west and north to the drier areas of the State, thus opening up the Western Plains and Peak Downs and leaving the coastal area for cattle grazing, sugar-cane and fruit farming. These inland flocks became highly successful wool producers and consisted of almost pure-bred merinos. Large sheep flocks were controlled from a head station, which had numerous peripheral outstations to form almost a complete small town in itself. Water obtained from bores formed an oasis of English flowers, vegetables and lush greenery around a timber-verandahed homestead, which kept at bay the arid heat of open, featureless plains.

The shearing of scattered, almost wild, sheep flocks on the fringes of outback settlement was carried out in portable sheds made from canvas sheets draped over demountable timber frames (in the early days, two poles sunk into the ground) and tied to the earth with heaped sods of earth. The early Queensland woolsheds were generally larger than those of New South Wales, and shearing was undertaken on a larger scale there than anywhere else in Australia. In many ways these runs were similar to those of Western Australia, sheep first being imported there in 1834 by a settler from England. Flocks were built up from the solid robust sires obtained from the South Australian and Riverina studs, and were grazed on land considered almost desert. These dry open areas contrasted strongly with the small-scale hilly country of Tasmania (or Van Diemen's Land as it was known in early years), where the landforms and colder climate were similar in many ways to parts of England.

Fine-woolled merino sheep first came to Tasmania in 1820 due to the influence of Lieutenant-Colonel Sorell, who obtained 300 merino lambs from Macarthur's flock. These sheep were distributed among the settlers around Hobart Town by the Governor, in proportions depending (in his opinion) on the trustworthiness and agricultural skill of the settlers. In 1825 the Van Diemen's Land Company was formed and their action proved an important factor in developing the sheep industry. An area of 325,000 acres of land was given to the Company under charter and, by 1830, it had 6,129 well-bred sheep, many purchased from Germany and George III's Spanish merino flock. Because of these importations, Tasmania was later to be called the 'Merino Stud of the World'. However, the island state was not to remain a grazing area for long. Larger estates were broken up to form smaller and yet smaller holdings and, in some cases, orchards were planted in their place. Sheep flocks, homesteads and woolsheds expressed this different scale of operation as a change of industry changed the way of life for many people.

In each state, as land was developed, the sturdy pioneers constructed their homes and work places from whatever the environment offered, in a strong, forthright way. These first structures were erected without fuss, were simple in design and fitted perfectly into the countryside from which they had been carved.

It was during the squatting times in Australia that some of the finest examples of indigenous woolsheds were built. This came about partly because the squatters' homesteads and farm buildings had to be erected in a hurry and partly because they were considered temporary structures, no great loss being incurred if law enforcement meant moving family, farm hands and stock elsewhere. Once the squatter was on the land, buildings were quickly begun. The site for a head station would first be chosen and casual labourers from neighbouring stations attracted to help in timber-splitting and building operations. Within about two days a couple of rough bark huts, consisting of no more than about 25 sheets, would be put up to serve as temporary

accommodation. Once the base was established the men set to work on permanent shelter. The roofs were of various types; sod earth was sometimes used, but usually grass thatching and wood shingles were favoured. Timber known as red beefwood or Botany Bay wood was sought after because it was easily split.

> A ten-foot by twelve-foot hut, with slab walls and a bark roof, could be erected quickly at a cost of £10: there would be one for the men and one, probably a trifle larger, for the master. Almost for a decade, the men in the bush lived in such erections of bark and slab, in a simplicity that reached the point of savagery.[3]

To ensure security for the boundary runs of the property, outbuildings were erected at strategic points about eight or ten miles away from the head station. The men in these groups of smaller huts ensured safety for the sheep flocks and they guarded the run. At the outstations lived a hut keeper and two shepherds, and hurdle yards of split palings enclosed flocks which were watched over by the hut keeper from his sentry box. A rough shed, consisting usually of no more than enough pole supports to hold up a bark or sod roof, was located near the men's shelter. This served as a woolshed and boasted a wool press and a flattened level space in the centre for the shearing of sheep.

The average squattage included little more than that. The rugged, locally hewn materials, constructed in as quick and as straightforward a way as possible, were the pioneers' statement of a tough, hard working life. It would have been insane to erect permanent or extensive spreads in a situation as insecure as this and, in the bush, rudimentary shelters that showed no signs of city luxuries were the best protection from bushrangers.

During this time, woolsheds as buildings in their own right became established. Many a selector or squatter considered the woolshed as being the most useful (and therefore the most important) building to him, apart from the homestead.

> The object most worthy of attention after this [i.e., the main hut or homestead] is the woolshed, first because it is the place where all the sheep are shorn and all the wool is sorted and pressed ready for carriage; secondly because, in magnitude, it is the largest building on the establishment.[4]

Woolsheds were often built first and the farmer would live in this shelter until he had completed his homestead and other farm buildings. From the erratic progress made by the first settlers in Port Jackson in using local materials, especially timbers, the settlers in the late 1850s could boast a wide knowledge of choice and use of these for building purposes. Gum, ironbarks and cedars were rapidly felled to make way for grazing lands, furniture and buildings, in many cases some of Australia's best timbers unfortunately being almost cut out within a single generation. The more rugged, simple and appealing structures, and certainly those using the most timber, were the animal and fodder enclosures. Where good timber was not available in large quantities and where other materials were easily handled, and less expensive, then stone, brick and, later, galvanized iron were used in considerable quantities.

During this time it was common for outbuildings to be constructed from slabs and logs—12 in. to 15 in. diameter logs being embedded in the earth, with adzed timber slabs horizontally or vertically fixed to a top and bottom plate forming the wall cladding. Scantling poles, still with bark sheathing, became rafters and smaller poles formed battens for wood shingles or thatching.

The grazier of squatting times had many other factors with which to contend. Larger flocks meant rationalized methods of handling the masses of sheep, moving them from the farthest paddocks, washing them, sheltering them from extremes of the environment prior to shearing, and enabling the animals to be moved through a series of races and pens for the shearer to remove the fleece in the quickest and most economic manner without damage. The sheep were then moved through a further series of races and pens, treated for various ailments and herded out to graze once more. The woolsheds were, therefore, large enclosed buildings (to accommodate many more stands) some hundreds of feet long, with floors of bracken and leaves to catch sheep droppings. Sheep still had to be washed before shearing began and the pioneer farmers developed their own variations on sheep washing. The most popular was that of damming the local stream or river and regulating the flow and volume of water by means of timber spouts. Close to the property would be a log and mud dam with two large box spouts or hollowed log spouts placed securely just below the water level (about two feet), with one end blocked with a slide door, drawn aside to allow the water to flow as required. Large frames of scantling poles formed into squares were built in the water below the spout openings. One sheep washer stood in each, with clothes and boots still on and in water up to his waist. A race led from a small yard close to the water's edge and, when the race was full, the sheep were given a hosing down and let stand for half an hour before they were plunged into the largest square in the water, then passed underneath the poles until each man had a sheep to wash. Once the fleece had been vigorously squeezed, the sheep was passed to the spout men, two men to each sheep, one taking the forelegs, the other the hind legs, and the sheep rolled to and fro until properly washed. When let go, the sheep swam to a pebble beach to drain and dry. The

shepherd then herded the clean, dry flocks along the less dusty tracks to grassed areas enclosed by hurdles less than a mile from the woolshed.

After the sheep had survived the wash pool it was necessary to keep them clean while they waited their turn for the shearer. To overcome the problems of exposure to temperature and climatic extremes, sheltered pen areas were built. The simplest type of shelter was one which was open-sided, using timber logs to form posts, beams and rafters, with bracken ferns as a roof. The other was a building similar in form to the woolshed, complete with enclosing walls.

> A long shed was erected from the entrance of the catching pens into the shed—split slats 18 in. in width, rounded posts 12 in. in width placed 3 ft. in the ground, adzed off each side; two hardwood battens nailed on the sides of the posts, slabs fitted close together one after the other up to the wall plate of the building. On each side of the shed a frame of hurdles was erected to serve as a loft. Bracken ferns were cut. . . . These lofts were also kept full; when the sheep were let go in the morning to graze a man forked up the bedding and put fresh down.[5]

Because sheep had to be quickly moved through the shed, the fleece had to be removed as rapidly as the bow shears would permit, and many more hands were employed to leave the shearers wholly to the task of shearing. No longer was it expected of him to wash the sheep, nor was it for him to doctor the cut sheep—tar boys were occupied full time to do this. A separated area was set aside at one end of the shed for the wool room, to which the pickers-up brought fleeces to be roughly classed on a timber bench and then stored away in timber-railed bins prior to pressing in hessian bales. A great number of fleeces filled the wool room and classers and pressers worked constantly, the latter being the brawniest of wool men who tightened a lever-arm in order to press as many fleeces as possible into each bale. Full, stitched bales received the stencilled insignia of the owner, property name and description of wool type, and were then loaded on to a bullock dray. Wool rooms, because of their greater volume of activity, grew to be areas separate from the board and were defined as such, usually by a timber screen wall or a higher floor level to be at a level closer to that of the dray deck. Early wool presses added great colour to this area, being timber panelled, carefully jointed and covered with coloured 'caravan' lettering bearing the makers' inscriptions. Many different presses were developed: some worked on the lever principle, others screw-wound or with lever ratchet catch.

Often the early wocl presses were box frames with vertically fixed slabs of timber forming sides, and were placed outside the shed. During the squatting times the wool press, as an essential part of the shed equipment, earned its position near the classing area and adjacent to a pick-up area for the bales. These bales found their way by various means to giant wool sorting rooms in the capital cities, before shipment overseas. Most common was for large, cumbersome loads of bales to be transported on creaking drays pulled by sturdy bullock teams. Camel trains were also used and, in later years, were a familiar sight in outback New South Wales, Queensland and South Australia. Properties fortunate enough to be sited beside a river used steamboat transport, with waterways as roads to larger towns. The inland river steamboats, with their enormous cargoes of wool, became commonplace as they steamed their way peacefully from rough outback stations to the developed larger settlements. Wool transportation for sales and shipment completed the last stage of a complex journey which had started on the sheep's back a matter of a few weeks previously.

The success of a wool clip in the squatting times depended on a host of operations performed by a great number of workers whose activities centred around the shearing shed. It was a golden decade, but in no way similar to the golden decade to follow.

WOOL BARGE ON THE DARLING.

The grazier's response to the challenge of providing satisfactory accommodation of differing types and creating physical bounds for large numbers of sheep (which, he quickly learned, behaved in manipulatable ways) had resulted in a distinct rural Australian architectural form by the late 1840s. However, his contribution did not stop there: he had also unwittingly created a new work-force to fill spaces built specifically for men and sheep—a highly organized factory space in many ways—and this work-force became more varied in time, as specialization became necessary for the increasing work load to be efficiently carried out. It was this specialization among the work-force that led to the rich, varied, descriptive phrases that were to become synonymous with Australia and the wool industry: each participant with his own jargon that he developed and added to; tall stories that abounded during spells of wet weather, when sheep could not be shorn, and in the evenings when the men gathered around the camp fires.

There were shepherds and gatekeepers to guard flocks, sheep washers, musterers, penners, shearers, bosses of the board, pickers-up, broomies, tar boys, wool classers, wool pressers, balers, dray drivers, shearers' cooks and rouseabouts. All earned their distinction during and after rural expansion of the squatting era in Australia and all made the woolshed a lively and rich centre of activity for the shearing months of the year.

Shearers began working in small groups and travelled from station to station as work presented itself. Although not properly organized, they followed a general thread of climate change from north to south over a thousand miles of eastern Australia, moving to each new shed as summer weather forced shearing to be carried out. The grazier anxiously prepared for this time by passing word around the local township that he needed a certain number of shearers and workers to carry out various tasks and, in later years, he provided rude but practical, simple shelter for the men. Most of the work-force arrived the day before shearing was to begin and each man was allotted a position on the board. Rouseabouts and tar boys were less skilled than shearers, and therefore not paid as much, but wool classers were relatively highly paid and considered the experts of the trade. Most shearers, if moving from station to station in groups, travelled on foot, humping their swags. They earned respect at all stations they visited on their way, where simple food would usually be given free or for minor services carried out around the property. This was not regarded as charity, but as a recognition of a fluid work-force that was, of necessity, to be maintained. The swagman was thus a product of the rural development in the days of expansion and settlement.

This attendant work-force of the industry was large and plentiful and squatters and settlers began to reap the benefits of years of toil. A stability of labour existed that satisfied the grazier and his employees.

However, this stability did not last long: in February 1851, close contact and knowledge of the land by a small group of people led to startled exclamations that were to resound as social upheaval was felt throughout Australia and the world—the cry of gold!

The first payable gold in Australia was discovered in the rugged country near Cornish Settlement (out from Bathurst) by William Tom Jnr, James Tom and John Lister, led by Edward Hargraves. The first three members of the group used their detailed local knowledge to apply Hargrave's advanced technique of gold panning methods, using a cradle to obtain specks of gold that were plentiful enough to be considered of payable quantity. William Tom named the site of the discovery, where the Lewis Ponds and Summerhill Creeks meet, Ophir.

Considering the millions of pounds' worth of alluvial gold that were subsequently dug from the river-bed and surrounding craggy hills, this was hardly a misnomer. However, the thousands of people who rushed to Ophir and to fields in Victoria (where gold was discovered during the same year) had a profound and deeply felt impact on the pastoral industry, and meant reorganization of property management and enormous changes in the rural building forms.

For the grazier the gold rush spelt disaster. Shepherds and convicts living on the fringes of grazing lands had discovered gold as early as the 1830s, while prospecting and fossicking in creek beds as a diversion and escape from the boredom caused by their lonely existence. Their finds were often not disclosed, nor allowed to gain undue attention, by quick intervention of the property owner who sensed an immediate loss of all he had laboured to create.

The gold finds near Bathurst in New South Wales and Ballarat and Bendigo in Victoria could hardly be ignored nor passed off as an isolated find. The farm work-force decreased rapidly in the first two years of the gold discoveries and in many cases became non-existent as servants, ticket-of-leave men, shepherds, rouseabouts and shearers all joined the feverish throng who downed tools and walked off to seek their fortunes.

For a number of years then, the grazier commanded vast acres of grazing land and swelling flocks of sheep almost single-handed. The first problem to be overcome was the protection of sheep flocks in outlying areas, which had normally been the concern of shepherds, and the only recourse was to construct hundreds of miles of timber railing to enclose the holdings. So the first fences began to appear and cut edges into large areas of bushland. Once flocks had been safely enclosed in paddocks delineated by fencing, it was then only necessary to employ boundary riders (many of whom were aborigines) whose task it was to repair broken fencing and report on the wellbeing of the flocks. The property was soon found manageable with a fraction of the previous staff, and more cheaply.

Finding shearers was more difficult. One woolgrower, who badly needed shearers for the annual shearing, in desperation sought assistance on the gold-fields and offered almost anything in return. The men agreed to do the job but, as payment, demanded all the wool shorn. When he rejected this the gold-diggers offered the woolgrower a job as cook at fifteen shillings a day! The only way the thousands of sheep could be moved through the sheds each year was by the graziers engaging shearers from other districts. Being for the first time aware that his skill was highly sought after, the shearer soon developed professional attitudes and in later years proudly fought for, and won, proper wages through group action.

The gold-fields offered success to very few. By the early 1850s the Australian population had rapidly swollen due to the influx of German, English, American and Chinese gold-diggers who had learned of the Australian finds or had moved from less successful diggings in California. With their dreams of success they brought a variety of skills, rich and varied cultural backgrounds and, in some instances, knowledge of the latest developments in industrialization. Chinese gold seekers, trekking overland from Robe, South Australia, to the Ballarat gold-fields, were forced to find work en route and were responsible for building a number of Victorian woolsheds. Some of the best outstation shepherds and cooks the woolgrowers employed were Chinese, too, their quiet and solitary nature making them well suited to spending weeks or months tending sheep flocks in the farthest fenced paddocks.

By the time alluvial runs had been worked out, and even the most optimistic gold-digger was looking for other employment, the grazier had attracted thousands willing to join the wool trade.

Gold rush times hastened the introduction of a revolutionary method of building construction that had developed in America and swept the continent: the system of balloon framing. Whereas some home-tried and perfected methods of housing construction were short-lived in Australia, it was the balloon frame system that caught on in almost the same degree of universality as it had in America. Machine-cut timber, wire nails and sheet iron, all the outcome of a takeover by machines produced by the Industrial Revolution, were the necessary basic materials. The previous construction methods, using heavy oversized logs, intricately jointed and locked by timber pegs or fixed by hand-made nails, gave way to the use of precision-cut rectangular sectioned timbers that only required butt jointing and banging together with factory-produced nails of standard lengths and shank sizes to provide quick and economic shelter.

This framing system applied to all elements of the building structure. Boards were nailed to joists to

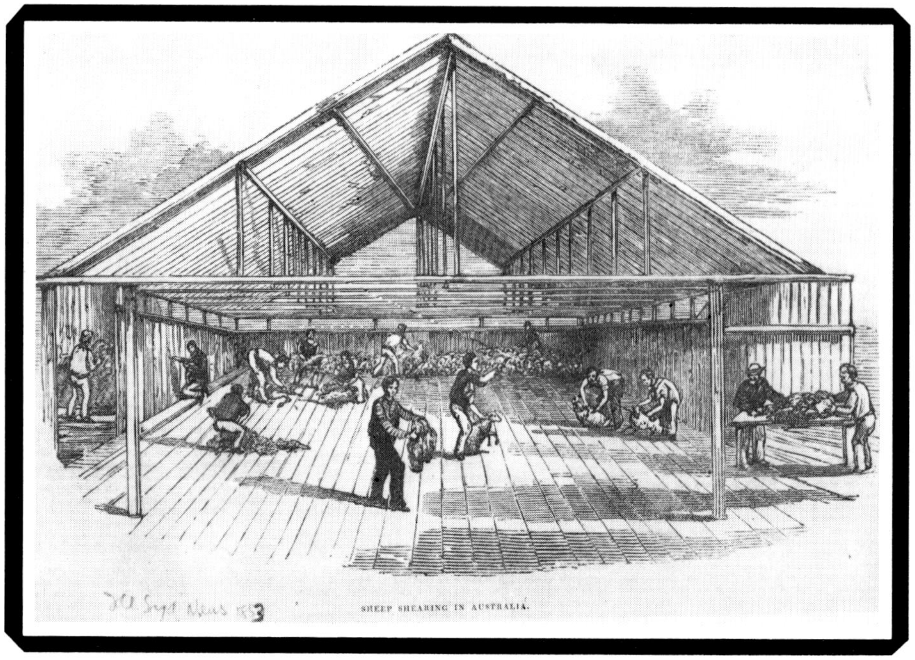

SHEEP SHEARING IN AUSTRALIA.

form flooring and walls consisted of studs regularly spaced, with nogging between, to form a frame which was prevented from swaying movement by bracing with timber of smaller section checked flush into the studs. It was found to be simple to prefabricate buildings and assemble them on site. The system of trusses and columns, as used throughout England in the nineteenth century for steel railway bridges, was utilized where it was necessary to provide large rectangularly-shaped plans. In this case, timber trusses were rapidly positioned on timber wall-plates fixed to stud framing and sheet iron fixed to purlins, which in turn spanned between each timber truss. This system was to change the whole character of Australian woolsheds.

Within a few years sheet iron, protected from rust by a process of galvanizing, was used in a great number of ways, not only for roofing but for the external cladding of farm buildings and where over-sensitive ideals did not prevail. When it was found that greater bending strength could be obtained from the material by stamping out the sheets in corrugations, and that this corrugated iron could be curved, water tanks and, in some cases, fences were constructed from it. Iron roofing, drums, cans and tanks soon were a necessary part of the grazier's work environment, and became a part of the indigenous materials at hand in the same way as timber had been previously.

It is important to note that at this time brickwork had begun to supplant timber in many buildings, mainly houses and public buildings in the cities and larger towns. But the advance of this permanent material was generally largely ignored when woolsheds and other farm buildings were built, brickwork usually being reserved for the homestead where finishes and prestige were considered of greater functional importance. The comparatively few woolsheds built of brick are found to be of early construction and often built by convict labour. Most storage, wheat and woolsheds were clad in weatherboards or iron sheets fixed to sawn timber framing as this could be constructed much more quickly, more easily and cheaply than brickwork, and rude practical detailing of joints between materials did not unduly worry the grazier. Corrugated iron was easily fitted to existing thatch, bark or wood-shingled roofs on older sheds.

With this almost universal acceptance of factory-produced building elements came problems inherent in the materials themselves. The most serious one was that of providing comfortable working conditions inside the new sheds for the skilled wool men. As the thin metallic cladding above their heads effectively conducted and radiated heat from the sun in the mid-afternoon of summer days, so it dripped moisture after cold mornings in winter. A minor problem in the coastal areas was the noise caused by sheets of rain that rattled and drummed on the iron roofing taut between light timber framing. As long as the sheets were properly lapped to prevent leaks, noise could be tolerated, but the shearers flatly refused to shear wet sheep, let alone shear in an over-leaky shed. Iron gutters and downpipes began to appear on many up-to-date sheds built during the 1850s and 1860s, giving the whole shed, with its sawn timbered, clean-edged, metal-roofed precision an almost finished touch of mastery over the elements.

Trial and error proved that iron roofs could be sloped at an angle that allowed condensation drips to run along the underside of sheets to the eaves rather than cause discomfort by dripping directly from roof slopes of a lower or greater pitch. Higher ceilings allowed hot air to rise, allowing relatively cooler air to be drawn from outside the shed. Methods of regulating ventilation varied. One popular method consisted of ingeniously constructed pivoted corrugated iron windows, or louvres, adjusted by means of a rope. This gave wall openings of variable size so that air was encouraged to move inside in gentle streams. Once roof ventilators (which stood as decorative elements on the roofs of many later sheds) became popular, the combination of air inlets in walls and outlets in high roofs solved many of the discomfort problems experienced in outback sheds.

Excessive damp in floors of sheds built close to the ground (which soon led to rot and decomposition of many fine timbers) was overcome by sitting the woolshed on higher log piers than normally constructed, thus allowing air to circulate under the floor. This also provided the solution to problems of sheltering mobs of sheep from exposure to climatic extremes as they waited their turn for the shearer's blade. Separate shelter areas were found unnecessary when the under-floor space of the shearing shed itself could be used. With the shed at a higher level, bales were more easily moved and loaded on to dray decks and, when the extra floor height meant dray decks were actually lower than the wool room floor,

FLEECE
1st PIECES
1st LOCKS

gravity was utilized to help move the rich, cumbersome cargo.

By the late 1860s, machine-sawn weatherboards were produced in large quantities and found their way overland from housing estates in the suburbs of richer cities and towns to newer farm buildings. Once positioned, lapped and fixed to studwork, this timber sheathing could be painted a variety of colours or simply left to weather a natural light brownish-grey in the sun. Weatherboards gave an added touch of prestige to the woolsheds and other farm buildings which had previously been clad in galvanized iron. Once stud walls had been boarded on the outside and clad on the inside, the occupants of these early buildings were spared the heat caused by radiation from the thin metal sheet walling.

Not all the newer woolsheds, however, accepted this influx of factory-made and machine-cut materials. A strong indigenous tradition of building in stonework existed in the Western District of Victoria, where basalt and granite are spread over thousands of square miles of the best Victorian sheep country. The basalt favoured for building construction, known as bluestone, lies thick on country lacking natural timber. (On Titanga property, Lismore, early pioneers scattered thousands of blue gum, sugar gum and other eucalyptus seeds over ploughed land in order to provide shelter for future sheep flocks.) Early use of bluestone in Victorian woolsheds may be seen at Eurambeen, Beaufort (*c.* 1845), where even, neatly fitting layers comprizing small and large bluestone blocks are laid on lime mortar. Carefully squared, the largest blocks are approximately 12 in. deep by 18 in. wide and 9 in. to 12 in. thick. Early bluestone sheds display simplicity in form and construction and are usually whitewashed internally. Decorative use of the material is evident in sheds of later 1880 boom times when architects were engaged to design complete property buildings.

Industrialized timber construction was also barely considered in parts of South Australia, where a tradition of construction had evolved through the availability of limestone and ironstone in large quantities. Mostly the stone blocks in these sheds were larger than those of Victorian bluestone and were random-stacked with very little lime mortar but squared carefully (for construction purposes to ensure straight length of walling) at corners. Many sheds displayed a double height section to accommodate the screw press, the overall building form with its combined one-storey and two-storey form expressing the different space uses inside.

By the 1870s the Australian woolshed had begun to lose some of its ruggedness, but what it had lost in the warmth of oversized rough-hewn local materials was gained in size, scale and precision of organized construction.

The woolshed displayed a new sophistication in design and, set apart from the homestead, it became one of a group of buildings. Close to it were grouped smaller units consisting of simple timber huts for the throng of shearers to inhabit. Shed hands occupied one or two huts separate from the shearers, and a cook, in many cases Chinese, was engaged full time either by the grazier or the men themselves to feed the workers and take the usual abuse levelled at the cooking. These men were, by the 1870s and 1880s, well organized as a labour force. Shearers moved from station to station, some on horseback, some in sulkies, others on foot. Many of them arrived on bicycles that were pedalled, ploughed or dragged wherever man could go. Their clothing and equipment became traditional (almost a uniform in later years), having developed from the necessities of work use and transport. Clothing had both to stand up to rugged wear and tear in the sheds and be light, portable and easily strapped to the bicycle along with the shearer's blanket, tea mug and handpiece. Proper footwear was usually reserved for pushing the bicycle over bumpy outback tracks and so, on arrival at the shed, the man would procure an old wool sack and fashion about a dozen pairs of moccasins which usually lasted him until the shed cut out.

In the big sheds there was a hum of activity which started with the morning bell and lasted until knock-off time at sundown, and there was much to interest the wool-gathering team—rivalry between shearers, bets among them as to who would be the ringer, and the rabble of disputes after figures showing the number of sheep shorn by each man were chalked up on the tally-board at the end of each day. It was a time when each shed had its ringer who swore he had shorn more sheep per day than anyone before, as the tally-board took no account of the difference in sheep and the cutting quality of wool in the various sheds where so-called exceptional tallies were made.

Although there were always disputes as to shearing records, some became universally recognized and therefore accepted as fact. As early as 1874 'Long' Maloney shore eleven big wethers in eleven minutes at Fowler's Gap. One sheep per minute was sometimes common when shearing lambs and one of the first accounts of such a record was in 1884 when Sid Ross, at Belalie on the Warrego River, shore nine lambs in nine minutes. A shearing match was staged at Parratoo, South Australia, in 1876 between Angus McInnes and Jack Gunn, and the sheep were big sandback wethers. However, there were not enough penned for a full day's shearing and Angus McInnes had the £50 prize-money after shearing only 180. The highest tally made with blade shears was in 1892 at Alice Downs station, Queensland, by Jack Howe, one of the greatest Australian shearers. Jack Howe's record was 321 sheep in eight hours. This time was often hotly argued as the day's run was eight hours and forty minutes. However, the other shearers had begun their day's work when Howe commenced his record run and he was back in the hut before they had finished. Howe's effort was hardly an isolated inspired effort for, in that same week, his average was an incredible 259 merinos per day, and he won added bonuses for cleanness of blow and care in handling the sheep. It took many years to beat this time, and then it was with the use of machines. In 1892, the same year that Jack Howe scored his amazing tally, a team of shearers at Northampton Downs, Queensland, averaged 200 sheep to a man and, at Barenya in 1895, eight men averaged 236 sheep each, the 26 shearers on the board averaging 175 sheep each in one day.

After working hours there were various activities to involve the shed hands and shearers—concerts, card games, foot races (another opportunity to place bets) and sometimes fights. The local publican usually earned the most from shearing time once a shed had been cut out, as the froth of beer flowed non-stop to help celebrate the completion of weeks of sweated labour. In parts of outback New South Wales and Queensland, a day's horse-racing was commonly organized at a local pub to mark the cut out of shearing at sheds in a district. Tambo, on the Barcoo River, was well-known for many a wild race meeting as big local sheds mustered large mobs of horses for the occasion (some fine gallopers, others pack-horses), and there was keen competition between shearers, squatters and locals. Among shed followers were some who were experienced jockeys, as well as being very good shearers. At cut out time there was a large cheque for shearing and a better one to be won at the races afterwards. After the merriment of these celebrations, however, there were those who went to the next shed literally penniless.

While shearers took wool off hundreds of thousands of sheep each year with blade shears of various types, and tallies of 200 per day were common, there were those who held that industrialized techniques were the means to achieving progress and, by 1871, several patents had been granted for sheep shearing machines. Attention was turned to these machines soon after the perfection of a horse-clipping machine in 1870. None of these early mechanized shears worked satisfactorily, but incentive to develop the idea was strong, especially where flocks were large and shearing was such a vital element of the pastoral industry.

On 28 March 1877, Frederick York Wolseley, who owned Euroka property near Walgett, New South Wales, was granted a patent described as 'Improvements in apparatus for shearing sheep and clipping horses and in contrivances for driving them', but it wasn't until 1885 that Wolseley, in partnership with mechanic John Howard, perfected the full apparatus. In this same year Wolseley held a public exhibition at Goldsborough & Company's wool store, Melbourne, to prove to all the future of his machine shearing device. Blade shearers shore in opposition to the machine and won but, on a second going-over, this time with the machine-run handpiece, the blade-shorn sheep yielded an extra two and one-quarter pounds of wool. The graziers, journalists and interested company men needed no further demonstration and thus the whole character of the Australian woolshed and the wool industry was to change.

While Howard travelled to the large sheds in Queensland and New South Wales to demonstrate the benefits of machine shearing to mistrustful shed men, at Bourke, New South Wales, a further industrialized advance had been made. At Toorale station, in July 1887, the first electric lights were installed in an Australian woolshed. This not only extended working hours but opened up possibilities for the installation of electrically run machinery.

A small group of sheds had applied for, and were completely fitted out with, machine shears for the 1888 shearing season. The Dunlop shed at Louth, near Bourke, New South Wales, was the first to have machine shears installed and was the testing-ground for success elsewhere in Australia. The Dunlop shearers' reactions to these innovations were almost predictable. After roll-call in July 1888 the 40 stands stood empty, with 184,000 sheep waiting to be shorn, as the indignant strikers retired across the Darling River and set up camp. Both sides stood mistrustful of the other. After a deadlock lasting nearly three weeks a decision was made that, at any cost, shearing was to begin. A cook was elected, stand numbers were drawn and the correct method of using the machine was demonstrated on several sheep. Before long, backs were bent and the fleeces rolled from sheep shorn by the first commercially set up machine shed in Australia.

The most up-to-date machinery that the industrialization of the 1880s could produce, almost incongruously at first, threaded its way through rugged timber log beams and columns of sheds which had seen no previous innovative signs since the two-wheeled push-bike. Steam engines, machines to drive flexible overhead shafts, overhead friction wheels and claw-like handpieces of various designs found their way into even the remotest of rudimentary outback woolsheds and were to become as familiar, and as much a part of the Australian woolshed, as the galvanized iron water-tanks that were propped on timber poles beside them.

Once the shearers had overcome their initial prejudice towards the industrialized shears, a new prosperity was carved off the sheep's backs and tallies soon greatly exceeded those of blade shearing times. At Barenya shed, Jimmy Power made a record by shearing 315 mixed sheep in eight hours with machine shears. Dan Cooper at Bundoran, Lower Flinders, in 1910 beat this record by one sheep, and also shore five sheep in the same number of minutes with machine shears. It was in the 1870s and 1880s, with this heightened interest in the industry and in competition, that a wave of prosperous, sophisticated and even architect-designed woolsheds began to appear, and property buildings reflected the creativity and wealth of boom times.

Most of the 'prosperity' sheds took account of space required for an engine to drive the machine shears and a room for the mechanic, or expert. This extra space was simply grafted on, in economically available materials, to sheds that had made the changeover to machines. In new sheds these rooms formed separate areas near the wool room, with access from one to the other and with the expert closest to the board so as to give immediate attention to faulty machinery.

Of the big sheds built in Australia (where numbers of sheep shorn each year were often as high as 200,000), many in the Riverina district of New South Wales employed between 50 and 100 shearers. In 1899 at Bowen Downs, Queensland, 364,742 sheep were shorn and this was for some time considered the biggest sheep station in the world. Brookong shed, near Lockhart, New South Wales (built *c.* 1900), had 97 stands and shore 7,000 to 10,000 sheep per day. There has been as much controversy over claims for the largest Australian woolshed as there has been over shearing records. Reputedly, the biggest shed was Tinnenburra, Queensland, with 101 shearing stands, but this claim is also made for Burrawang, New South Wales, both sheds now unfortunately destroyed. Burrawang utilized two Fowler steam engines which were brought overland from Adelaide to drive the machine shears and two wool presses. Classers worked above the presses and fleeces arrived to be classed via a conveyor belt. The shearing board was so long and there were so many men working there that two overseers were employed, one at each end of the board. Legend has it that a person could get the sack at one end and find a new job at the other. Burrawang one year scored a massive clip of 4,994 bales and, in 1902, during drought conditions, was able to supply the nearby township of Ootha with fresh meat.

Yet, despite the build-up of prosperous runs and a fully employed work-force, sheds were burnt to the ground by shearers in Queensland during the 1890s. In February 1891, organized pastoral workers and employers clashed in what was almost civil war over wage claims, working conditions and employment of non-union labour. On many properties gates were opened, linch-pins removed from wagon wheels and threats of violence made. At Nive Downs the woolshed was burnt to the ground along with 150 bales of wool. Police and troops arrived and were reinforced by armed stockmen. Violence reigned in Augathella. After employers met at Barcaldine, further unrest flared as non-union labour was employed to replace striking unionists. This resulted in unionists leaving open all the gates between Bowen Downs and Aramac, a distance of about 45 miles. Further attempts were made to burn down sheds near Muttaburra though, fortunately, these were not successful owing to the intervention of a squadron of mounted infantry.

After men had drifted back to work and the sparks of unrest between the Australian Workers' Union and the Graziers' Associations became fewer, changes came about in woolshed labour organization, and arbitration through the courts was set up at the beginning of the twentieth century. The powerful Australian Workers' Union, in 1907, established through the court a standard of wages, hours of work and conditions for the shearing industry. Under the contract system a contractor organized teams to shear the graziers' sheep and the graziers were relieved of the work of dealing individually with the men. Huge companies were formed to undertake this work on a nationwide basis. However, the grazier soon found he had no voice in the running of the shearing. In later years, contract shearing was abandoned in favour of co-operative shearing companies. Notable among these is The Graziers' Co-operative Shearing Co. Ltd., better known as Grazcos. The new system enabled Grazcos, by the later 1930s, to extend its organization throughout Queensland, New South Wales, Victoria and Tasmania, with the shearing of over seven million sheep each year. A new stability in the industry had been found.

During the times of industrial unrest and, in some districts, the added toll of drought conditions, the Australian woolshed had undergone further changes. It emerged as a consciously designed building that both suffered and benefited from advances made in new agricultural and scientific methods. Funny Hill, Binda, New South Wales (built 1906), shows a fully matured indigenous style with its considered arrangement of shearers' huts and other outbuildings, consistently bonded together and given character through use of timber framing and weatherboarding. Such examples of scale and uniformity are rare in the buildings of today. Closer settlement, brought about by land division acts and the Government scheme of Soldier Settlement (breaking up of properties to accommodate ex-Servicemen) robbed the homestead and its farm buildings of this scale, character and functional importance.

The big sheds were almost to vanish from the far west owing to lower sheep-carrying capacity, in turn caused by lack of suitable stock feed, rabbit plagues and the memory of past scorching droughts. The woolshed structure lost more and more of its body as skeletal hardwood trusses and sawn timber rails gave way to finer, spidery steel members. By the war years and later a whole individualized and distinctive Australian architectural style was almost extinguished or built out, lost to the impersonality of characterless materials, and to sheds of a scale and atmosphere that were never to match the exuberant masculinity of those former years. Many a shed hand in the 1890s complained that things weren't the same since the blades were no longer used—and in many ways he was right.

Wayne McPhee, 1972.

[1] Keatinge, M., *Travels Through France and Spain to Morocco.*
[2] Colwell, J., *The Story of Australia.*
[3] Roberts, S. H., *The Squatting Age in Australia 1835-47.*
[4] Beatty, Bill, *A Treasury of Australian Folk Tales and Traditions.*
[5] Lawson, R. A. S., *Umbiwana.*

Planning Arrangements.
Although dozens of different types of plan arrangement for sheds are to be found throughout Australia, three of these, basically, capture the essence of the woolshed.

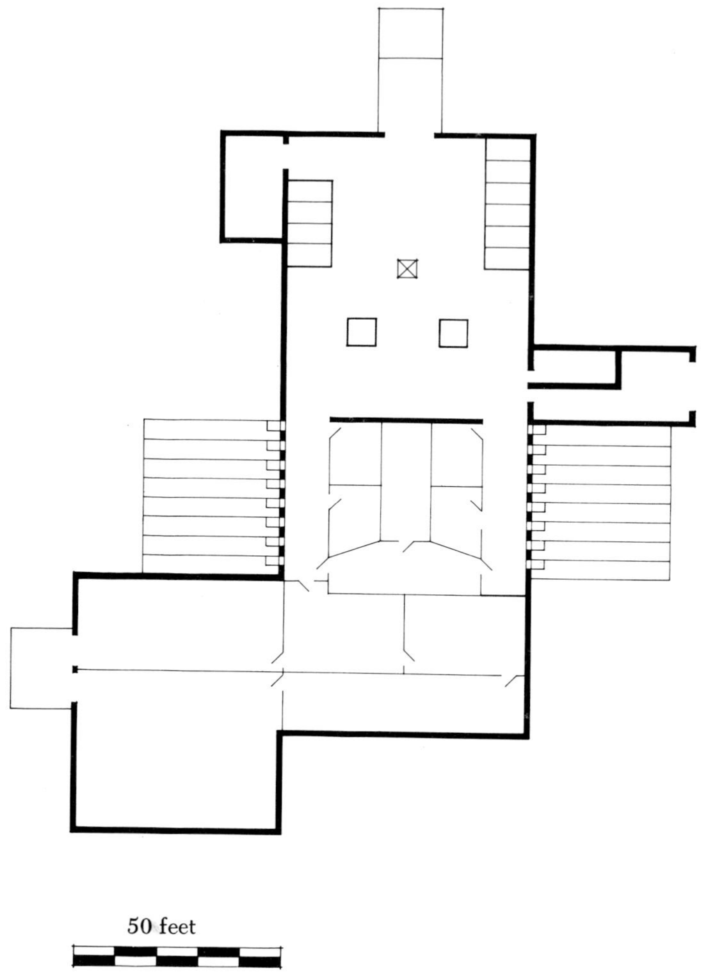

Sheep are herded up a ramp into large sweating pens, then turned into the smaller catching pens in the centre of the shed. The shearing board runs either side of the catching pens and, once shorn, the sheep are dispatched down a ramp into counting pens. The wool room is a screened separate area containing classing tables, sorting bins, wool press and storage space for finished bales. Office, Expert's and engine rooms open off the wool room.

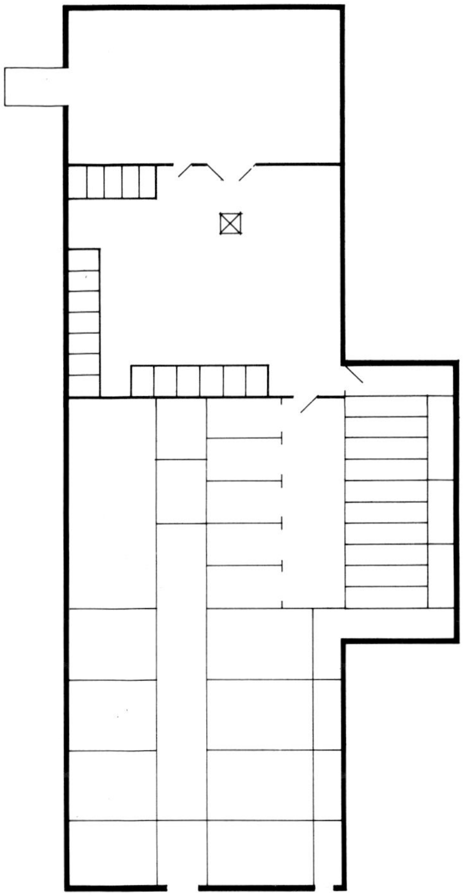

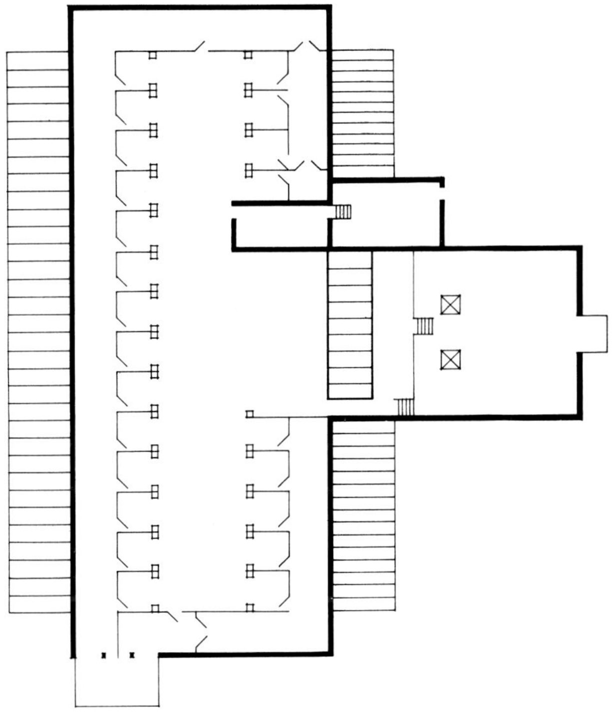

The second arrangement is for movement into one end of the shed and into pens in the centre. The shearing board is placed between catching and counting pens. Sheep, having been shorn, move out parallel to their inward journey. At the opposite end of the shed is the wool room, extended to provide a separate wool storage area.

Another planning variation, similar to that used in many of the large sheds. Sheep are brought into the shed via a ramped entrance and are dispersed around the perimeter. Shearers work along two sides of a centre board, moving sheep out through a chute to the counting pens. The wool room is divided from the main body of the shed by the classing bins, with the wool press at a lower level. Expert's and engine rooms are linked and open off the board.

Clayton Farm

CLAYTON FARM, Bordertown, South Australia

This building typifies early pioneer woolshed architecture.

Whole tree trunks are used as post supports to carry the trimmed-off branches and twigs used as rafters and thatching.

A building mound which has grown from the materials offered by the countryside, its timbers bleached like bone by the blistering South Australian sunlight.

Hollowed log feed troughs rest on the cobbled surround of the vertically fixed split slab walls.

Inside the shed the floor is compact earth except for an area around the wool press, where sacking is laid down.

Present owner: Miss W. Wiese.

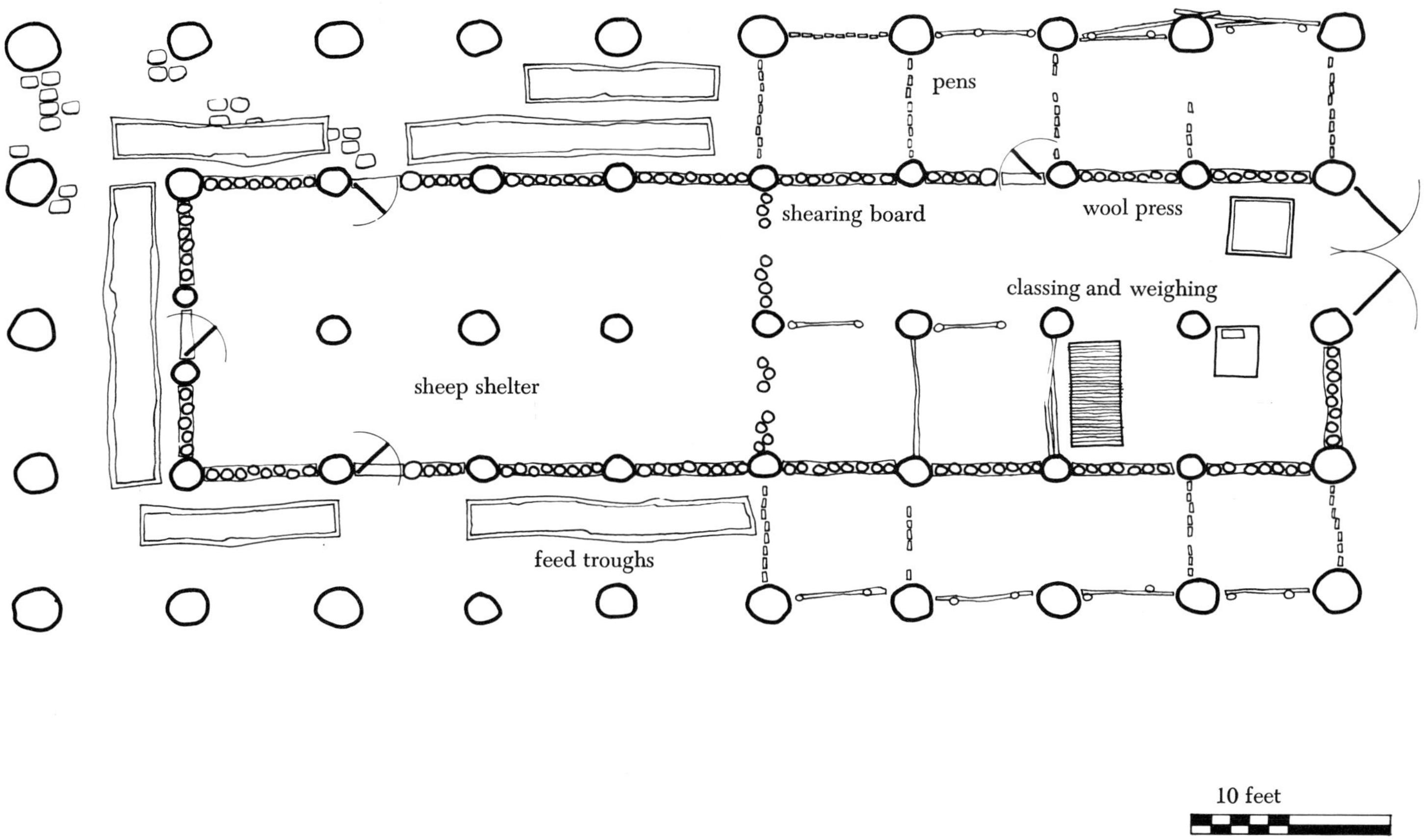

A mound that has grown out of the earth

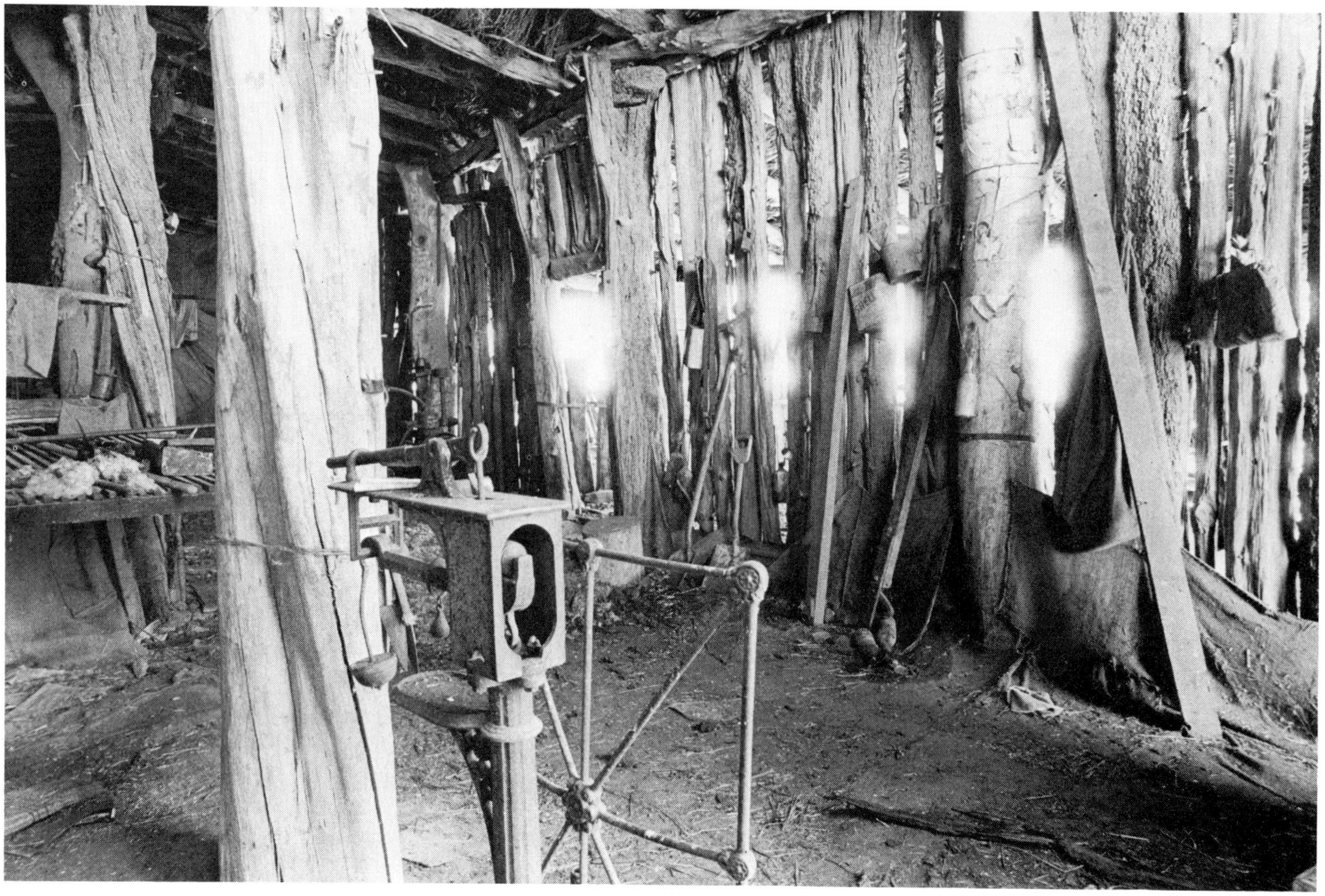

The shearing board

The wool press

Hessian lined sheep shelter

Door frame detail . . .

. . . and sheep pens

Sheep country in New England district of New South Wales

These two woolsheds on the Ophir road, near Bathurst, New South Wales, show typical small property sheds. Patched over the years, their exterior has developed a colour scale of browns, from the orange rust of the iron to the muted grey-brown of the timber.
The buildings hug the ground, their profile following the contours of the undulating countryside.

ROYALLA, Queanbeyan, New South Wales

A combination of weatherboard and galvanized iron clads a framework of logs.
This 4-stand shed has been given a new façade of aluminium-framed windows along the length of the shearing board, contrasting with the finely beaded sash windows of old.
Now it faces the danger of being swallowed up in the ever-spreading development around Australia's capital, Canberra.

Timber, glass and iron fuse together to make this strong pattern of portholes and counting pens

Wool bales awaiting transport, Markdale, Binda, New South Wales

Wool laid out to dry, Wingadee, Coonamble, New South Wales

NEWSTEAD, Elsmore, New South Wales

Tom Roberts, who has been referred to as the most significant artist in Australia, found in woolsheds a subject that reflected his love for the Australian bush. Newstead was one of the sheds that he painted and it was during his stay on the property with his friend Duncan Anderson that he painted his best-known work, 'Bailed Up'.

The small shed immortalized by Roberts is still in use and stands on Newstead North, owned by A. C. & E. Bucknell, Newstead South being owned by descendants of Duncan Anderson. About 1888 the Newstead South woolshed was built, using robust ironbark to form a structure of vertical split log infill between framing timbers left in the round. The shed is raised off the ground to a height of up to 5 ft. and is supported by approximately 240 timber piers. These timbers have been bleached by the sun and scoured by the wind to an elephant-skin texture.

Newstead North

Newstead South

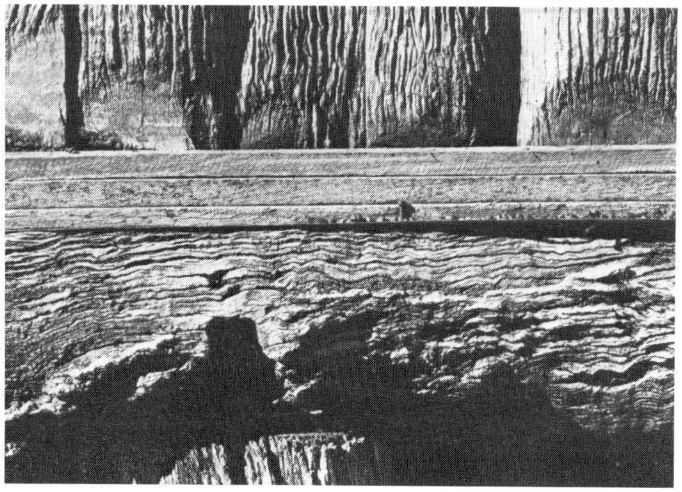

Timbers scoured by the wind

Newstead South

The woolshed seen from a shearer's bed, Funny Hill

FUNNY HILL, Binda, New South Wales

A fine example of a weatherboard shed, preserved and cared for over the years by the Carr family, who have owned the property since 1850.

It is built mainly of stringybark and yellow box felled and milled on the property.

The large ironbark posts (measuring up to 68 in. diameter) were placed in position with the aid of a bullock team.

This shed, built in 1906, replaces the original one that was destroyed by fire.

The old Ferrier press, No. 246, was saved and is still in use.

Present owner: James H. B. Carr and family.

Classers' tables and shearing board

The wool room

Stencils adorn the posts like votive offerings on a Greek icon

Where the shearer washes . . .

. . . and works

Shearing at Tottington, Victori

The woolshed has an atmosphere like no other building, brought to life once a year during shearing when it takes on a new lease of life.
The interior, dark by contrast with the fierce glare of the sun outside, has matured to a mellow brown. Its timbers have been burnished by the grease from the wool to a fine dull patina that is reflected by the sun burning through the skylights and doorways.
Men work silently and the sound is the drone of the shearing gear as the wheels spin above the shearers' heads. The shearers in their uniform dress of blue—singlet and denims—line the board, bent double peeling the fleece from the struggling sheep. The handpiece of the shearing machine buries itself again and again in the thick wool until the fleece is shed like an old coat.
No one works harder than the shearer; eight hours a day of back-breaking, sweat-drenching work. 7.30 a.m. till 5.30 p.m. This day is punctuated by the sound of the bell that brings relief from the shearer's monotonous task and splits his work time into four two-hour periods. 7.30-9.30, then a break affectionately known as smoko; 10-12; an hour for lunch; 1-3, smoko again, then 3.30 to 5.30.
As each man finishes a sheep and sends it scuttling through the porthole and into the counting pens, he painfully straightens up, pulls the cord that gives a short respite to his machine, walks the two paces to his catching pen, pausing momentarily to wipe the sweat from his face before selecting another sheep that he lifts and drags to the board.
Meanwhile, the picker-up has rushed forward to bundle up the fleece in his arms and dash it to the wool table. The rouseabout has swept the board clean and the aching process starts again.
The rouseabout must also keep the catching pens refilled by drafting sheep from the sweating pens. On the wool table the fleece is skirted before it goes to the classer, the wool expert on whose decision rests the grade and, eventually, the price of the wool.
It is from the classer's bins that the presser—stripped to the waist, his body oiled with sweat—takes the wool. It is then pushed, pressed, jumped on and cajoled into the ornate sentinel of the shed—the wool press. It takes a strong man to pump the lever that compresses approximately 400 lb. of greasy wool into a bale. Stretched plump and clipped, branded by stencil with property name, wool class and bale number, it is ready for transportation to the city wool sales.
Once again the shed is cut out and silent, yet somehow still inhabited by the ghosts of shearers past.

Pressing the wool, Oakwood, Queensland

The wool press, that point in the woolshed where the wool ceases to resemble a torn mattress and takes on the shape of a tidy hessian block. Today's wool handling techniques have produced both vertical and horizontal hydraulic presses but things haven't always been as easy as that.

> Adjacent to the woolshed was usually a rough wool press of split stuff—what Sir Edward Parry called a barbarous device, consisting primarily of a lever forced down by man's strength. To screw the bales was unforgivable, because it was thought to cake the contents (and, moreover, a screw press cost £60) and the squatter was satisfied with his rudimentary pressing. The fleeces were packed in the German mode, spread one on the other, the neck of the second on the tail of the first, and pressed into bales 9 ft. by 4 ft.—the sagging bales were sewn and tied on to the bullock wagons.
> from *The Squatting Age in Australia 1835-47*, S. H. Roberts.

In spite of this the screw press was developed and is still in use in Tasmania to this day. Typical of this type of press is the one (shown overleaf) at Earlham, Tasmania, made on the property. The hand-carved thread of the screw is made of the hardwood notelaea, and pit-sawn blue gum is used for the box.
In 1861 the Geelong firm of Humble & Nicholson started to manufacture the famed Ferrier lever press and the wool press became a highly ornate and richly decorated item in the otherwise rudimentary woolshed. To work this machine, wool is taken from any one classing bin and bundled into the press, the bottom box containing the open bale. The same action is used for filling the top box, which is then swung over into a position above the other. Then by winding, and finally by lever, this double filling is rammed into the single bale below.
Humble & Nicholson built over three thousand of these presses, and they are all entered in the firm's 'press book' in fine longhand. The oldest of these presses still giving useful service is probably the one at Mooleric, Victoria—this is No. 9. (In the same shed, early straight-spoked Wolseley shearing gear can be seen.)

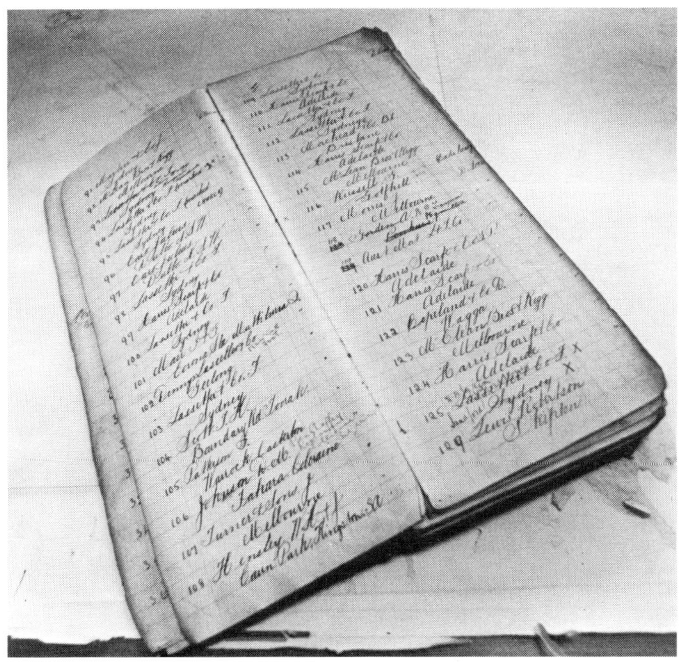

The press book

Lever press, Springfield, New South Wales

Screw press, Earlham, Tasmania

Woolshed interior, near Molong

This shed near Molong, New South Wales, shows how the outward appearance has changed through the use of different cladding.
Originally built of horizontal and vertical logs, at some stage of its life it took on weatherboard ends. . . .

. . . . Now that its life is ending, the roof has been stripped off to reveal the bones of the building, and shows the typical woolshed section that was to appear again and again in various forms.

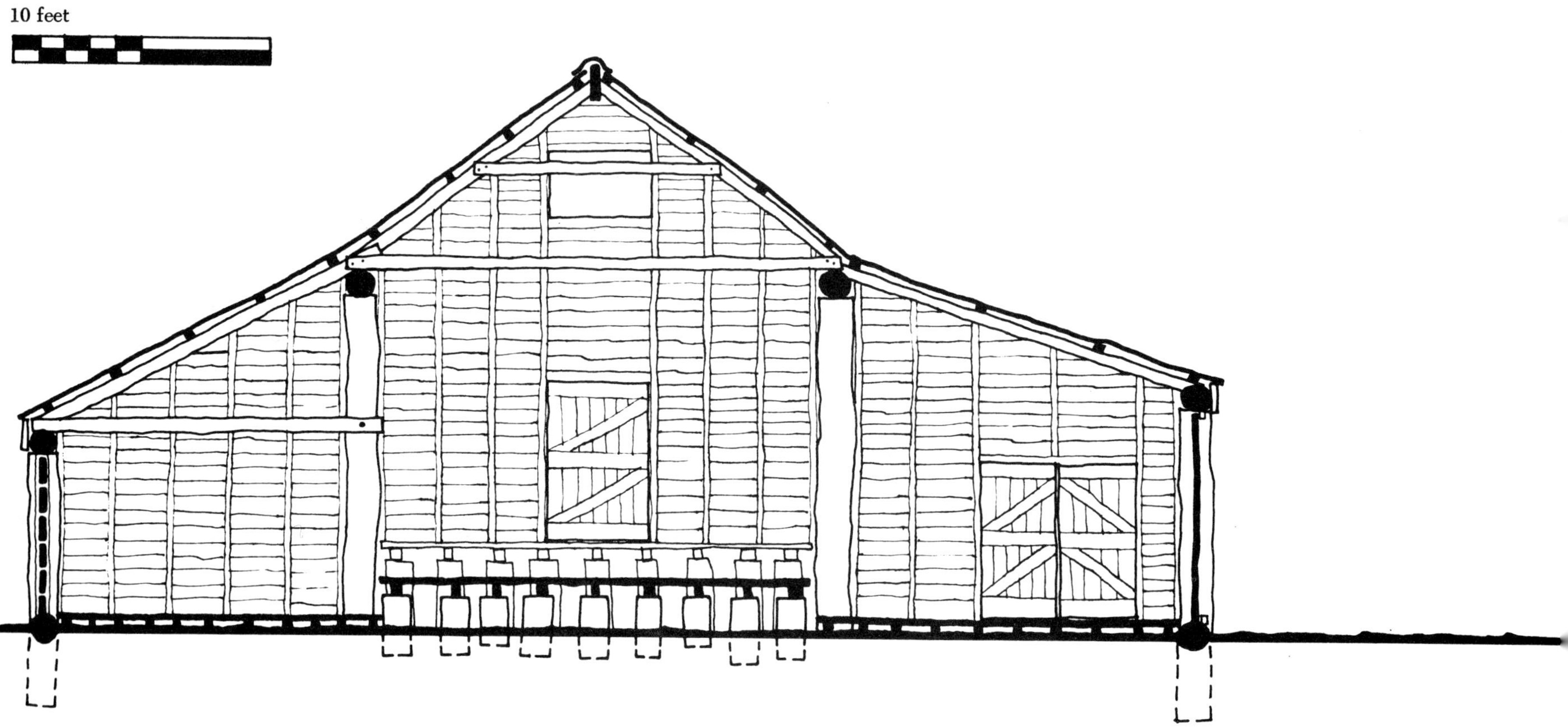

OLD URANGELINE, Lockhart, New South Wales

Bush carpentry at its best.
Construction generally is of 12 in. diameter logs left in the round for posts at 9 ft. centres, and the same for floor piers.
Floor joists and top-wall plates are similar logs, but adze-flattened on two sides.
Rafters of cypress pine are left in the round at 1 ft. 8 in. centres, with sawn battens supporting the original cedar shingles under a later cladding of corrugated iron.
The floor is of 2 in. round timbers laid on the joists at 2½ in. centres leaving a gap between each for self-cleaning.
The shed is approximately 240 ft. long by 60 ft. wide and is said to have contained 100 stands.
During the late 1890s up to 260,000 sheep were shorn annually at Urangeline woolshed.
Times and needs have changed and a smaller shed of galvanized iron has recently been added to the old shed, which is now used as holding pens during shearing
Present owner: Mr & Mrs Trethowan.

Winderadeen, Collector, New South Wales

Lake George from Winderadeen

Errowan Park, Carcoar, New South Wales, a galvanized cluster that makes full use of the topography

Wool chutes lead down to . . .

. . . the wool room

Shearing board, Errowan Park

Interior and exterior alike use strong no-nonsense bush carpentry detailing in this small shed near Marulan, New South Wales.
Present owner: Mr & Mrs R. T. Welch.

Stockyards outside Egelabra woolshed

EGELABRA, Warren, New South Wales

This woolshed is said to have been built by Chinese prospectors seeking work on their way through to the gold-fields.
Marks on posts and beams are likely to be Chinese symbols of construction details.
Egelabra boasts a stock of pure Spanish merino blood, Reg. flock No. 32, founded in 1897.
Present owner: H. E. Kater & Son Pty. Ltd.

Midkin stripped of its shearing gear

MIDKIN, Moree, New South Wales

This shed, that once throbbed to the sound of 32 shearers, now stands silent and empty.
The shearing board ran along two perimeter walls of the shed, with a central sheep pen.
An early use of machine-sawn timber can be seen in light sectioned framing members and roof trusses.
The property has now been split up and it seems that the fate of the shed is demolition.

The impressive scale of Toganmain

TOGANMAIN, Hay, New South Wales

The Toganmain woolshed has the very essence of legend and lore. It is one of the famous sheds in a chain of properties in the rich Riverina sheep country.

Built prior to 1876 by ship-builders of the sailing ship era, it had a centre board of 82 blade shearers.

In 1887 the Wolseley shearing machine was demonstrated during the shearing period and the first 100+ to be shorn in one day by machine was recorded there.

The following year 15 machine stands were installed at a cost of £120 and by 1911 this had grown to 55 stands. Today only 21 are still in use.

Samuel Shumack, in *An Autobiography or Tales and Legends of Canberra Pioneers*, writes:

> About 1876 Duncan Cameron and James Murty were shearing at Togglemayne Station, in the Hay District, when they met the brothers Ned and Dan Kelly. Dan was tar boy and Ned was a shearer. These friends of mine declared that Ned and Dan were well-conducted youths, and Murty said that during his forty years as a shearer he never met a nicer lad than Dan Kelly. They worked together for several seasons. Later, when the police party were shot at Stringybark Creek in Victoria, they were shocked to learn that Ned and Dan Kelly were the culprits.

However, although three Kellys were listed as shearers in 1876-8, the present owner, John Robertson, feels it unlikely that Ned the notorious bushranger was one of them.

Toganmain, having a frontage to the Murrumbidgee River, made full use of the wool barges that used to ply its waters. A horse tramway conveyed the wool to the wharf where it was loaded or, if the river were not navigable, stored until barges arrived, finally making its way to the Melbourne sale rooms.

Present owner: J. S. Robertson & Son.

Age-old interior of Toganmain with its portable 2-stand shearing gear

CLIEFDEN, Mandurama, New South Wales

Another woolshed with bushranger associations.
In 1862 Ben Hall and his gang raided the property and left their mark in the shape of two bullet holes in a wooden shutter in the shed.
After this, armed guards were positioned in the shed to safeguard the shearers' pay of gold.
Two hundred thousand bricks were made on the property and used to build the shed in 1842.
The interior is in three equal sections containing stables, wool room and classing tables. The shearing board and pens take the form of a low timber addition.
Present owner: The Rothery family.

1842

Echoes of the past can be heard in the wind that blows through this deserted shed near Bomera, New South Wales

The new image Gostwyck

GOSTWYCK, Uralla, New South Wales

The original Gostwyck holding owned by Henry and Grace Dangar, who came to Australia in 1821, was recently divided into two 6,000 acre properties, each owned by a great-grand-daughter of the original Dangars.
The homestead, stables etc. remain as Gostwyck, but the old woolshed is now on Deeargee land. This made it necessary to build, in 1968, the new 10-stand shed shown here.
The timber-framed building rests on concrete supports and open web steel joists replace the heavy timber construction used in earlier buildings.
It is probably the most up-to-date and best appointed woolshed in the area.
Present owner: Mrs N. N. Wright.

DEEARGEE, Uralla, New South Wales

The name Deeargee was taken from the original Gostwyck wool brand $\frac{\text{DR}}{\text{G}}$ after the division of the property.
This woolshed was built in 1851 by William Dangar, who settled the property in 1834.
The shed plan is a 50 ft. diameter octagon with a 75 ft. long tail. The octagonal part was originally used by 24 blade shearers working in a circle to take advantage of the good light that the roof form offered.
Now this section has been transformed into well-planned and well-constructed sheep pens that can house 1,800 sheep in full wool.
The shearing board of 10 machine stands is now in the tail of the shed.
The distinctive roof of the octagon is in three layers of galvanized iron separated by a clerestory of glass and ventilators, the supporting posts shaped to simulate the roof.
Vertical timber boarding makes up the exterior of the shed and brick was used for the wool room that was added a few years later.
Present owner: Mr & Mrs W. G. Giblin and family.

The shearing board, present . . .

. . . and past

'Wool away!'

I've been many years a shearer, and fancied I could shear,
I've shore for Rouse of Guntawang and always missed the spear;
I've shore for Nicholas Bayley, and I declare to you
That on his pure merinos I could always struggle through.

Chorus

But, oh! my, I never saw before
The way we had to knuckle down at Goorianawa.

I've been shearing down the Bogan, as far as Dandaloo;
For good old Reid of Tabretong I've often cut a few.
Haddon Rig and Quambone, and even Wingadee,
I could close my shears at six o'clock with a quiet century.

I've been shearing on the Goulburn side and down at Douglas Park,
Where every day 'twas 'Wool Away!' and Toby did his work.
I've shore for General Stewart, whose tomb is on the Mount,
And the sprees I've had with Scrammy Jack are more than I can count.

I've shore for John McMaster down on Rockedgial Creek,
And I could always dish him up with thirty score a week.
I've shore at Terramungamine and on the Talbragar
And I ran McDermott for the cobbler when we shore at Buckingbar.

I've been shearing at Eugowra, I'll never forget the name,
Where Gardiner robbed the escort that from the Lachlan came.
I've shore for Bob Fitzgerald down at the Dabee Rocks,
McPhillamy of Charlton and Mister Henry Cox.

That was in the good old days—you might have heard them say
How Skellycorn from Bathurst rode to Sydney in a day.
But now I'm broken-mouthed and my shearing's at an end,
And though they called me Whalebone, I was never known to bend.

Last Chorus

But spare me flamin' days, I never saw before
The way we had to knuckle down at Goorianawa.

Anonymous

WIDGEGOARA JOE

I'm only a backblocks shearer, as easily can be seen,
I've shore in almost every shed in the plains of the Riverine.
I've shore in most of the famous sheds, I've seen big tallies done,
But somehow or other, I don't know why, I never became a gun.

Chorus

Hurrah, my boys, my shears are set, I feel both fit and well;
Tomorrow will find me at my pen, when the gaffer rings the bell,
With Haydon's patent thumb-guards fixed and both my blades pulled back,
Tomorrow I go with my siding blow for a century or the sack.

I've opened up the windpipe straight, I've opened behind the ear;
I've practised every possible style in which a man can shear;
I've studied all the cuts and drives of the famous men I've met,
But I've never succeeded in plastering up those three little figures yet.

As the boss walked down this morning, I saw him stare at me,
For I'd mastered Moran's great shoulder-cut, as he could plainly see,
But I've another surprise for him that'll give his nerves a shock—
Tomorrow he'll find that I have mastered Pierce's rang-tang block.

If I succeed as I hope to do, then I intend to shear
At the Wagga demonstration which is held there every year.
And there I'll lower the colours, the colours of Mitchell & Co.
Instead of Deeming, you will hear of Widgegoara Joe!

Traditional

CLICK GO THE SHEARS

Down by the pen the old shearer stands,
Grasping the shears in his thin bony hands,
Fixed is his gaze on a bare-bellied yo
If he gets another one, oh Lord, won't he blow!

Chorus

Click go the shears, boys, Click! Click! Click!
Wide is his blow, and his hands move quick.
The ringer looks around and is beaten by a blow,
And curses the old codger with the bare-bellied yo.

In the middle of the floor in his cane-bottom chair
Sits the boss of the board with his eyes everywhere.
Notes well each fleece as it comes to the screen,
Paying strict attention that it's taken off clean.

The tar boy is there, waiting in demand,
With his blacken'd tar pot and his tarry hands.
See one old sheep with a cut upon its back,
Here's what he's waiting for, it's 'Tar, here, Jack!'

The Colonial Experience man he is there of course
With his shiny leggings, just off his horse,
Looks round the shed like a real connoisseur,
Brilliantine and scented soap and smelling like a whore.

Shearing is all over and we've all got our cheques,
So it's roll up your swags boys, we're off on the track.
The first pub we come to we'll all have a spree,
And everyone who comes along, it's come and drink with me.

Down by the bar the old shearer stands,
Grasping his glass in his thin bony hands.
Fixed is his gaze on a green painted keg,
Glory, he'll get down on it, ere he stirs a leg.

There we leave him standing, shouting for all hands,
While all around him every bludger stands,
His eyes are on the keg which now is lowering fast,
He drinks hard, he works hard, and goes to Hell at last.

Traditional

but the click of the shears
was to give way to a new sound in 1888
when the hum of machines was heard on the board.

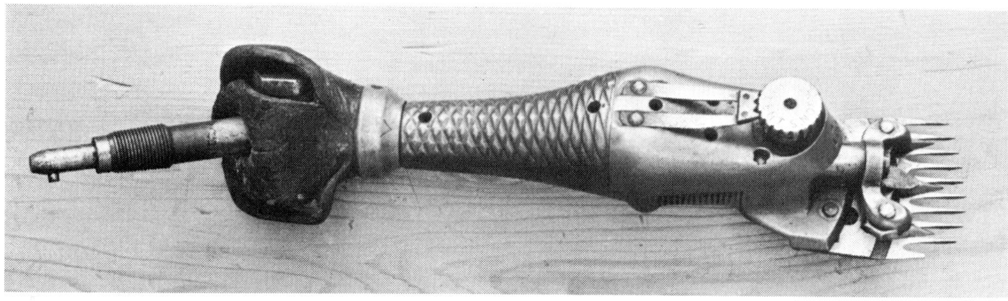

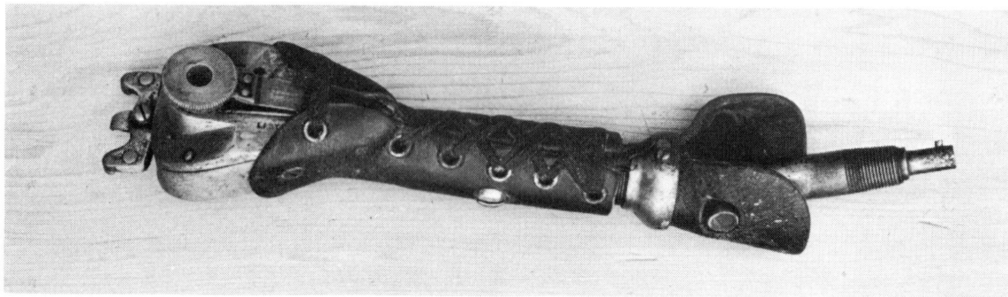

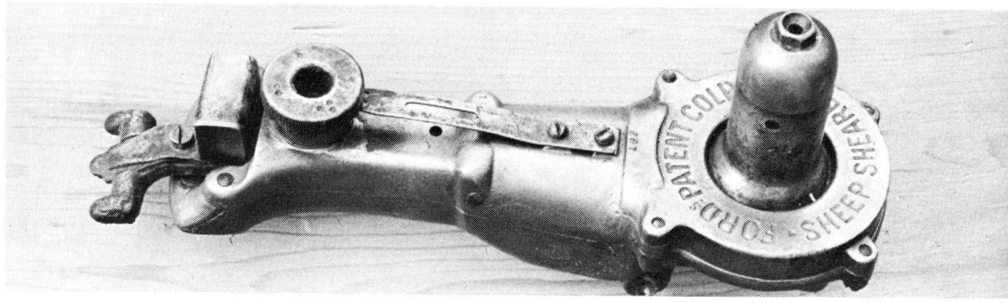
FORD'S PATENT COLD SHEEP SHEAR

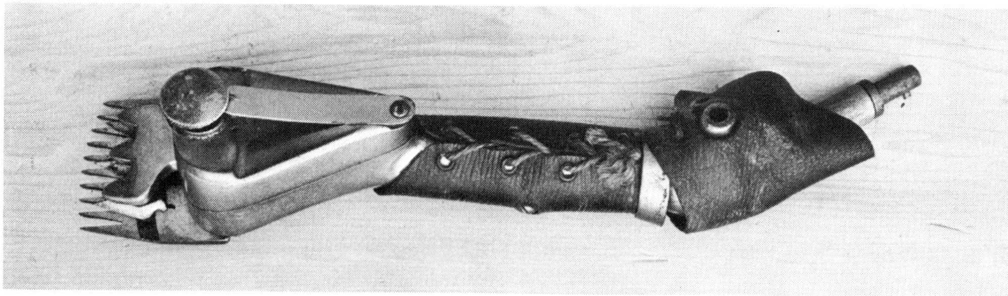

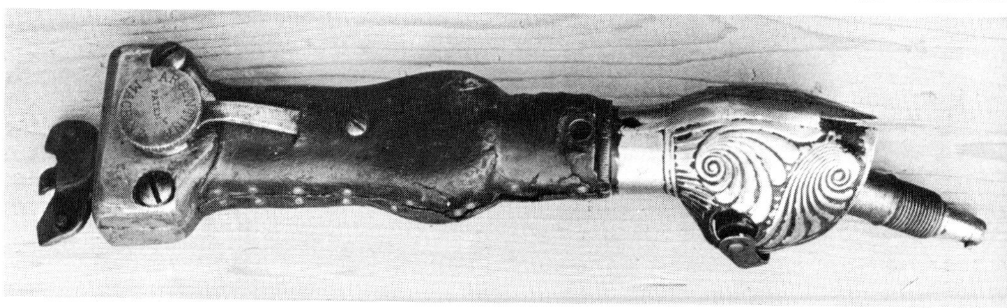

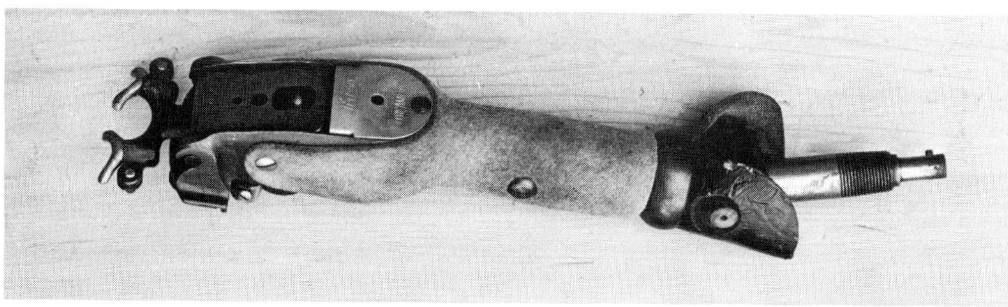

SMOKO

SMOKO

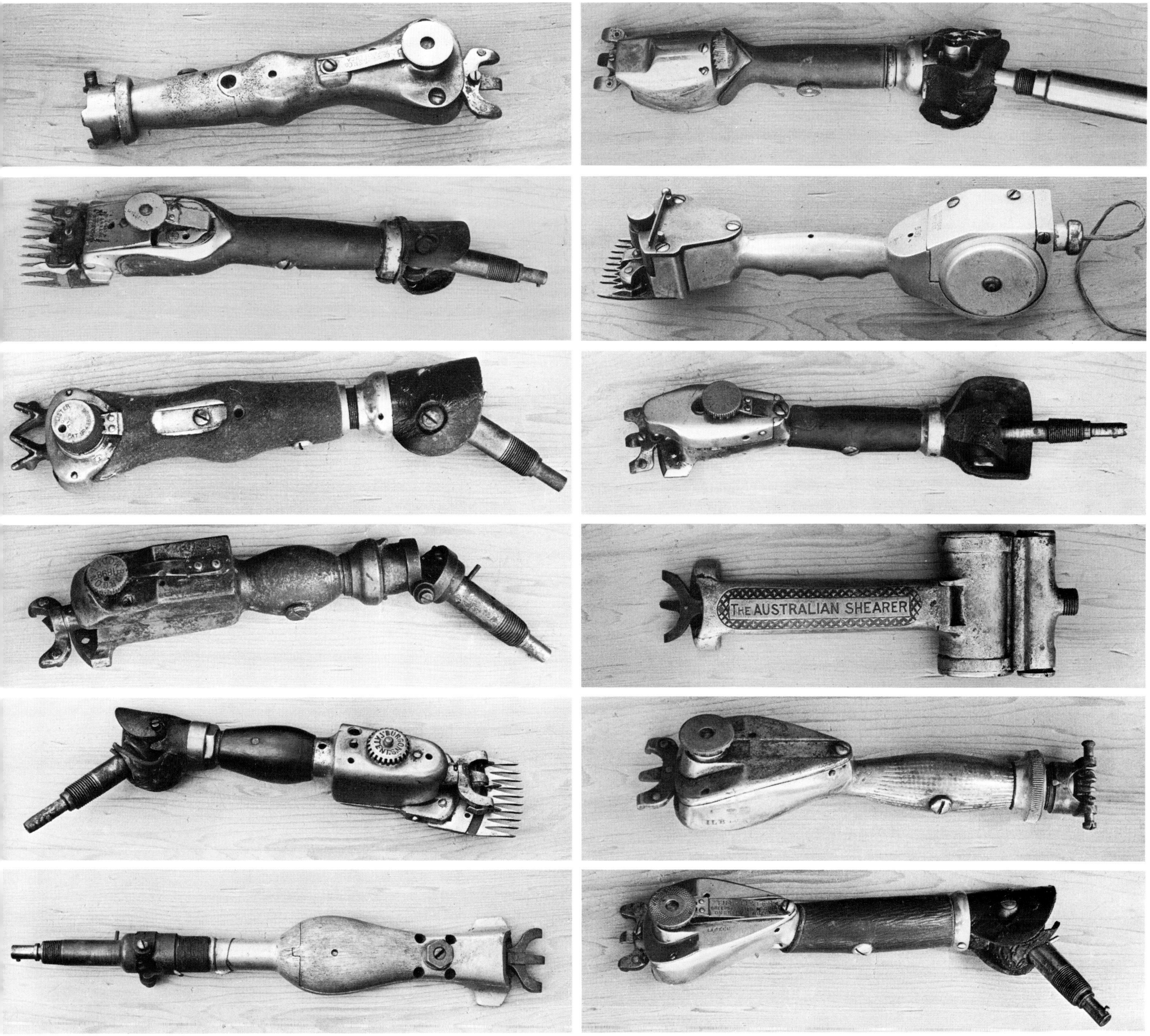

Early machine handpieces collected by Dangar, Gedye & Malloch and now in the possession of the Museum of Agricultural Progress and Rural Technology, Orange, New South Wales

PENDARVES, Byng, New South Wales

The original hipped roof section of this shed, built by convicts using local handmade bricks, has been added to in various stages in slab timber and galvanized iron.
The interior has been given a consistency by the use of white paint throughout, its history recorded in graffiti.
Pendarves property buildings were built in 1840 by George Hawke, one of the original settlers in the area.
Present owner: A. Hawke.

At Springfield the shearers' mess room is within the shed

SPRINGFIELD, Byng, New South Wales

Springfield was originally owned by 'Parson' Tom, a staunch Methodist, who along with the Hawke, Glasson and Lane families (all from Cornwall) formed Cornish Settlement.
Later mining success earned it the title Byng, and Springfield built up an extensive sheep run.
The woolshed's low pitched skillions merge into a steeply pitched central roof area, giving the shed a strong moulded form in the landscape.
Present owner: Mr & Mrs Kouvelis.

Rows of roof saplings . . .

. . . are mirrored in the wool bins . . .

. . . at Ghoolendaadi, Boggabri, New South Wales

Warragoon, Deniliquin, New South Wales

These two woolsheds show the universal use of galvanized iron, for roofing, wall cladding, water-tanks, even a chimney.
Used vertically and horizontally, timber poles of various sizes form races and pens.
A complementary use of two of Australia's best-known building materials.

Near St Arnaud, Victoria

Flattened tar drums used again

Bare bones of a shed near Wellington, New South Wales

Windswept shed near Lake Cargelligo, New South Wales

Timbers and patterns, Belltrees

BELLTREES, Scone, New South Wales

Built in 1875 of stringybark posts and beams clad in galvanized iron.
This shed, once for 40 blade shearers, now employs six machine shearers, and is one of the few sheds to boast two wool presses.
The light filtering through from the roof ventilators enhances the character of the slatted overhead catwalks and the open rails of the wool bins.
Present owner: H. E. A. & V. White.

Old homestead and new woolshed, New South Wales

Improvization is the building material of the poor.
Many a sheet of iron throughout Australia has been taken and used time and time again.
This woolshed near Lake George, New South Wales, is a typical example of man adapting this material to his needs.

Reevesdale

REEVESDALE, Bungonia, New South Wales

Not all buildings used as woolsheds were built for that purpose.
Reevesdale's was built as stables in 1829 and has since been converted into a 3-stand shed.
The two-storey brick building has a roof of pit-sawn ironbark clad with galvanized iron.
Internal posts of teak were brought out from India as ship's ballast.
One side of the nearby homestead has a row of windows painted over brick.
This was to deceive bushrangers and aborigines, who would fire guns and hurl spears at what they thought were normal window openings.
Present owner: Mrs G. Hurst.

The Shearer's Quarters
That group of buildings adjacent to the woolshed where he eats, sleeps, drinks, washes and sews, and spins his yarns of 'I remember. . . .'
This mess room at Woodside, Warren, New South Wales, shows the shearer's basic needs:
The lino-topped table and rough wooden bench.
Two refrigerators—the beer must be cold.
A meat hook in the ceiling and the alarm clock.

HADDON

HADDON RIG, Warren, New South Wales

The famed Merino stud of Haddon Rig was founded in 1882.
Always closely concerned with the standard of breeding, in 1953 a stud ram fetched the record price of 4,600 guineas.
The weatherboard woolshed was reconstructed in the 1930s during a vigorous improvement programme, and contains 58 stands.

Sundown

Windy

WINDY, Quirindi, New South Wales

A king among woolsheds.
This magnificent shed with its silver-painted exterior radiates in the sunlight, the silver colour giving a wholeness to the different materials of the 'wedding-cake' structure, tier upon tier.
The play of light upon horizontal and vertical, solid and void, give this shed a bold and vigorous character.
A hundred men were employed in 1901 to build this shed of 44 stands. All the timber, apart from the imported oregon rafters, is cypress pine cut and milled on the property.
Present owner: Australian Agricultural Co.

Sun and shadow

Tasmania's smaller scale land shapes repeat in woolshed forms

MARCHWIEL, Bream Creek, Tasmania

A family woolshed.
The property is still owned by the Dunbabin family who originally came from Wales in 1831. The family almost always had enough sons on the property to carry out shearing operations themselves and it was rare to employ outside hands. Today it is becoming more difficult, and wool is sent away to be classed and pressed.
Built in 1880, the woolshed proves the dictum that 'form follows function'.
The square wool room was built to a width and height that enabled the men who worked the screw press to walk around the press on a gantry above it, compressing the wool as they went.
Next came the shearers' board and classing table. This section was built to a man's height and included windows where light was needed. A piano case, used to transport the family instrument on its long sea journey from Wales, acts as a wool bin.
Finally, the area occupied by the sheep—the pens. It naturally followed that these should be sheep height and under cover.
Materials used for building the shed were those found on the property, stringybark and swamp gum, split and pit-sawn.
Newspapers (still bearing the date 1883) lined the walls to stop the wool from sticking.
Shingles replaced in 1918 still cover the roof. They were put in place by Mr James Dunbabin, father of the present owner, and he still considers it 'his' shed, as it was built only a short while before he was born.
Present owner: J. Dunbabin.

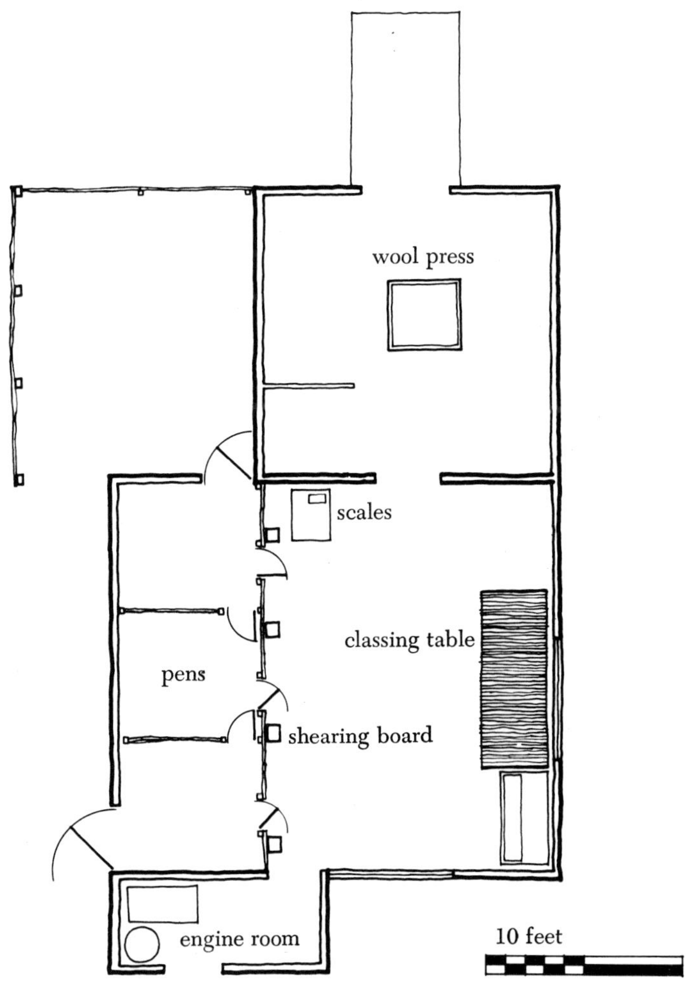

Functional grouping

Dungrove

DUNGROVE, Bothwell, Tasmania

This group of buildings has a strong unit identity.
The woolshed, stables and store shed have a common profile and character that binds them together.
Galvanized iron now covers the original shingles on the roofs, but the 1906 weatherboard is well preserved by paint.
The woolshed that still uses 8 of its 12 stands plays an important part in this handsome group.
Present owner: I. K. Downie.

Mountford

MOUNTFORD, Longford, Tasmania

Built around 1850 as a flour mill, though never used as such.
A complex of white-painted brick and corrugated iron.
The projecting roof over the storage loft and the fusion of angles and planes ally this building to the pure functional tradition of English mills.
In 1888 a flock of Corriedale sheep was founded, the oldest in the world, and the mill took on a new identity as a woolshed.
Present owner: R. A. Mackinnon.

Rudimentary detailing, Coolmore

COOLMORE, Nile, Tasmania

Iron and timber have weathered alike to an ash-grey in this woolshed, reputedly haunted by a shed hand who dropped dead during shearing many years ago.
Bush poles and split palings make up the walls.
Up to fifteen years ago, cut reeds laid on the earth formed a floor, and two men were employed during shearing time to cut and lay fresh reeds day by day.
Present owner: Boyes Bros.

LAWRENNY PTY. LTD., Ouse, Tasmania

The horizontal stripes of the weatherboarding emphasize the 300 ft. length of this shed.
High windows light a centre board originally designed in 1901 for 20 blade shearers.
Built on one of the oldest properties in Tasmania, which was settled by Lieutenant Edward Lord, Commandant of Hobart Town in 1810.
After the Second World War, in which three of the four owner-brothers were killed, the surviving brother, H. E. Brock, offered the property to the War Service Land Settlement.
In 1955 a company was formed by property owners to purchase the shed and facilities from the Closer Settlement Board, and it is now operated on a co-operative basis.

LOVELY BANKS, near Jericho, Tasmania

This small square sandstone shed of 1842 has been extended by weatherboard on a timber frame to provide a 4-stand shearing board.
The stone section of the shed used to house a screw press in the old days when seven bales a day was hard work, and it is most likely one of Tasmania's oldest woolsheds.

From the sheep's back, to bins, to bales

Raised slatted floor for self cleaning . . .

. . . and sheep shelter, Dennistoun

DENNISTOUN, Bothwell, Tasmania

White-painted weatherboard (again reminiscent of the mills of south-east England) give this woolshed a prominence and vigour in its landscape.
Set high above the ground on cut sandstone pillars and timber posts, it not only provides undercover shelter for sheep but is easy to clean.
The variety and placing of the openings complement the strong horizontals of the weatherboard and rainwater pipe.
Present owner: G. & G. Edgell.

PANSHANGER, Longford, Tasmania

The old and the new.
The 1821 woolshed was built by convict labour with brick walls and a shingle roof (since covered with iron).
Windows were barred to keep out marauding aborigines and bushrangers.
The present owner, wishing to improve on techniques of sheep-handling, built the new shed in 1950.
This shed, with its factory-like profile, is well and evenly lit by clerestory windows.
Ease of work for the shearer has been considered as the basis of this production-line approach.
Present owner: M. D. Mills.

Patchwork contours, Meadsfield

MEADSFIELD, Bothwell, Tasmania

Built just before the Depression, this shed was at the time considered to be one of the most modern in the state.
The timber and iron construction is levelled by concrete piers to compensate for the undulating land.
Present owner: John Lawrence Pty. Ltd.

WOOLMERS, Longford, Tasmania

'England expects that every man this day will do his duty   Trafalgar   Admiral Nelson   Duke of Bronte Oct 21 1805' is written on an overhead beam just inside this huge weatherboard shed, built in 1819 of hand-sawn peppermint gum secured by hand-made nails.
The shed is two storeys high in part, with the hand-made screw press standing proud and dominant in the centre of the top floor.
Such large timber buildings are unusual because of their high fire risk.
The old fire badge, Life Insurance Cornwall, is still over the door.
The property is still owned by descendants of the original settler, Thomas Archer.
Present owner: T. E. C. Archer.

Titanga

TITANGA, Lismore, Victoria

Built in 1872, when the Western District of Victoria was enjoying high wool prices, and property owners were employing architects to design their homesteads and other buildings, Titanga woolshed and homestead were designed by the Geelong architects, Davidson and Henderson, who were responsible for many other Western District property buildings.

The original fourteen handwritten pages of the specification for the woolshed is headed 'Specification of work and materials required in the erection of a proposed woolshed at Titanga, near Lismore, for Alex Buchanan Esq. in accordance with drawings numbered 1 to 3 and accompanying General Conditions and form of Agreement', and calls for the whole work to be completed in three and a half months.

Present owner: Partnership including Dr P. S. & Miss H. M. Lang.

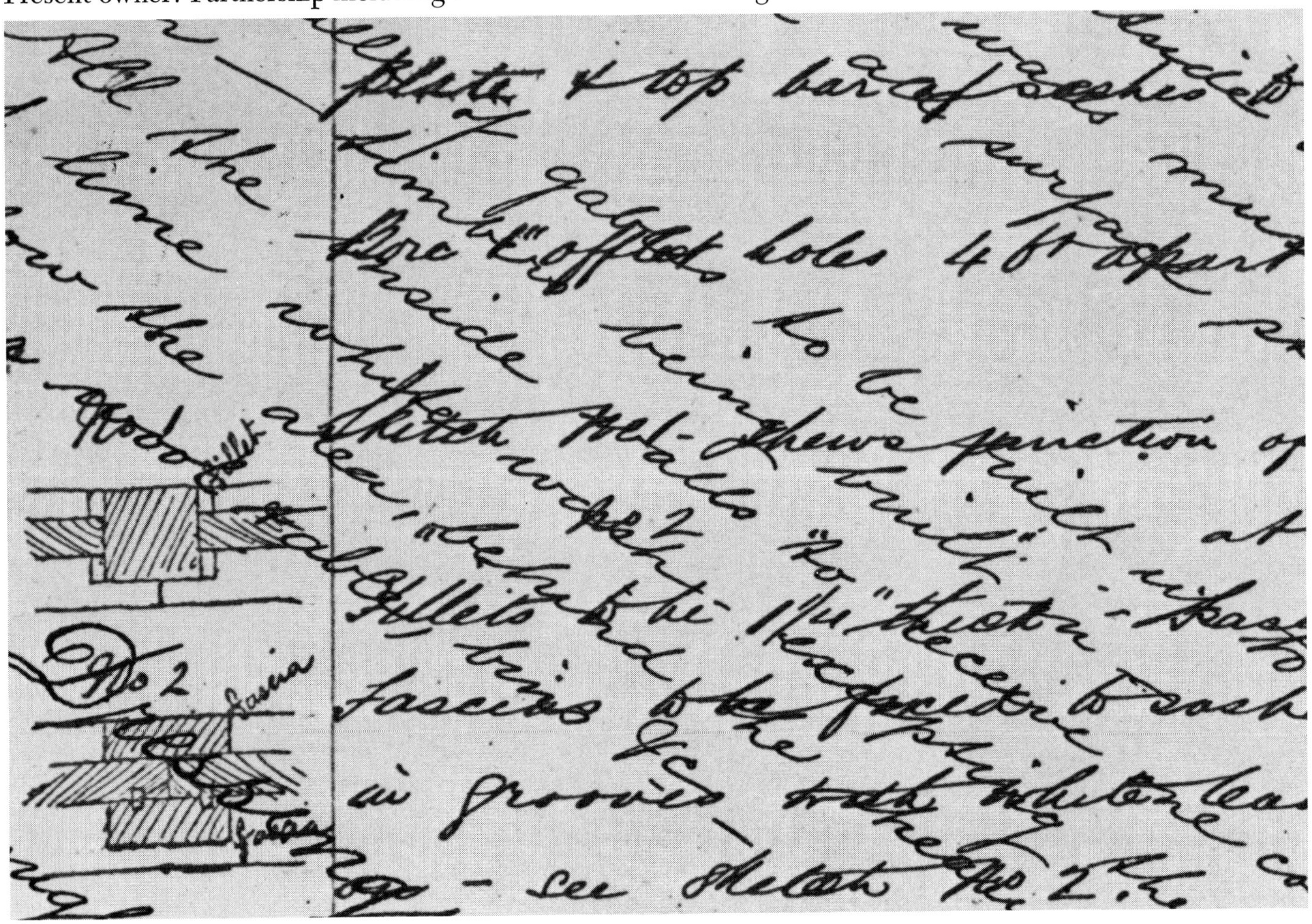

HUMBLE AND SONS
MAKERS GEELONG
1263
FERRIER'S
LEVER WOOL PRESS
1263
I
TITANGA

Shearer-controlled blinds shade the board

Hip and valley roof framing details, Eurambeen

EURAMBEEN, Beaufort, Victoria

Though no longer in use, this woolshed built in 1845 offers a great richness of texture in its robust bluestone walls.
The roof, still bearing its original shingles, also shows a forthright use of the materials that the nature of the property had to offer.
Present owner: Estate T. Beggs.

Barton, Moyston, Victoria

Terinallum, Darlington, Victoria

Mooleric, Birregurra, Victoria

Danedite

DANEDITE, Camperdown, Victoria

Surrounded by its accessories of water-tanks, well housing, pump and sheep dip, this fine shed sits proudly on its bluestone plinth.
Painted weatherboard with fine window detailing is used for external cladding.
Although not running sheep since 1948, the owners have kept the shed in first-class condition against the day when it may again come to life.
Two massive mess tables stand scrubbed and waiting for the shearers' return, the top of each made up of two pine boards 15 ft. 2 in. long by 19 in. wide.

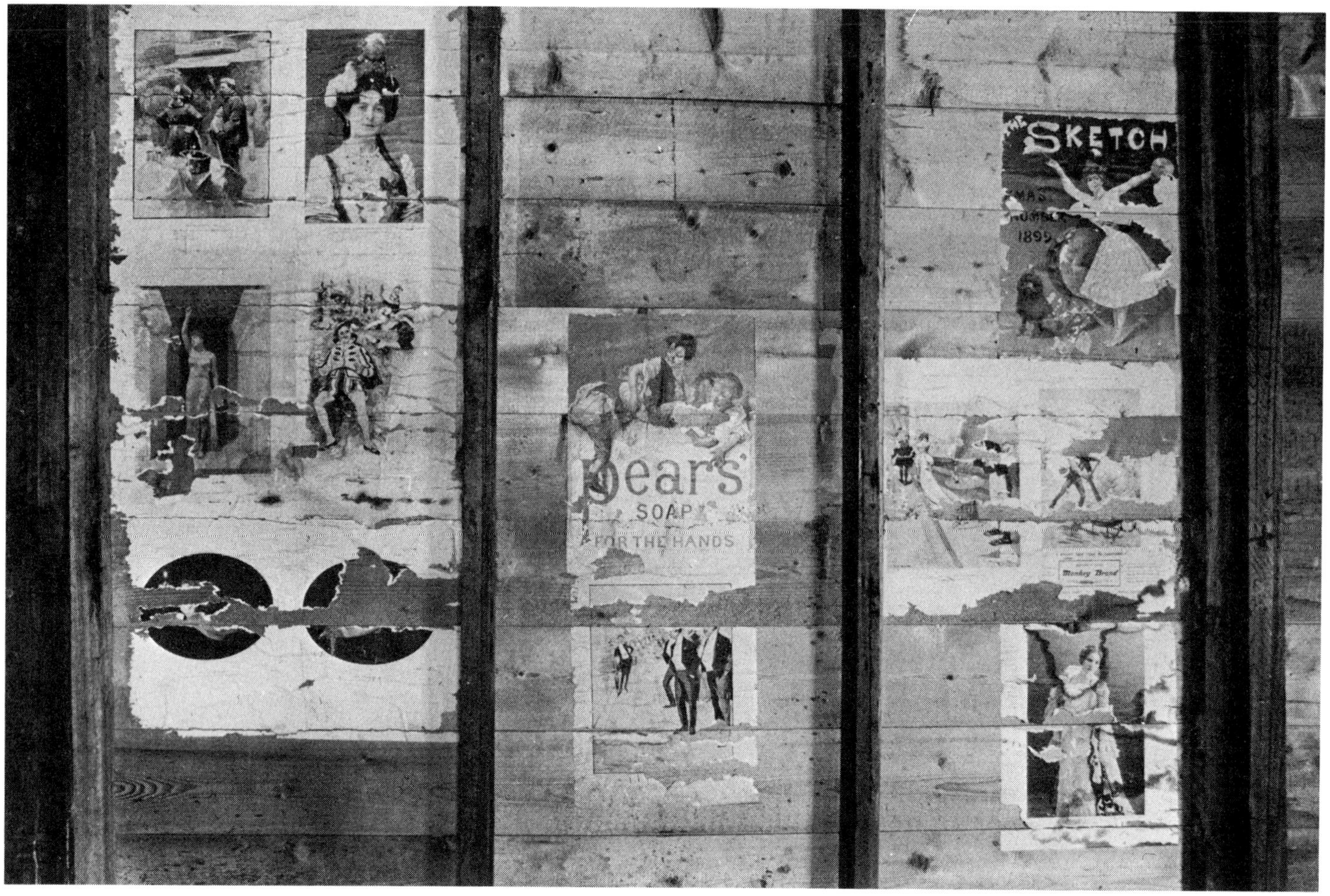

Pin-up girls of yesteryear . . .

. . . and stencilled reminders

Counter-balanced pen gates

Galvanized forms at Barunah Plains

BARUNAH PLAINS, Hesse, Victoria

A consciously arranged relationship of woolshed, shearers' quarters and cookhouse, well sited alongside the creek.
The shed of the traditional T plan uses bluestone, weatherboard and corrugated iron as each material suits. Six machines now take the place of the 20 blades on the board; this is lit and ventilated by high windows.
Present owner: G. K. Russell.

EAST LODDON, Serpentine, Victoria

Built at a cost of £9,000 in 1869, the plans being drawn up by the engineer-squatter, John Ettershank. His plans for the shed and a proposed grandiose house were elaborately drawn and painstakingly colour-washed, and are still in the hands of the present owners.
To build the shed and pave its yards 460,000 bricks were made and baked on the property.
The repetition of buttresses gives the 183 ft. long building a vigorous geometrical pattern of brickwork and louvred windows.
80 ft. long oregon beams support the roof of galvanized iron, and kauri pine is used for the floor.
Once housing 28 stands, today's conditions have reduced its use to 4 stands only.
Present owner: W. H. Long.

An engineer's forethought shows in this use of brickwork

The paved yards at Merrang

MERRANG, Hexham, Victoria

This woolshed was built around 1865 by Robert Hood, grandfather of the present owner, during his second extensive building campaign after taking over the property in 1856.
The first bluestone building has been extended over the years to provide undercover pens. The raised floor of this extension gives a strange scale to what was an external arched opening.
Another feature of the shed is the extensive use of stone cobbles in the outside yards.
Present owner: Robert Hood.

GORINN, Ararat, and GNARPURT, Lismore, Victoria

Two fine examples of bluestone woolsheds.
The forms of the galvanized roofs on their stone walls sit well in their woodland surroundings.
At Gnarpurt the strong arched doorways are emphasized by the white-painted timber doors. During the First World War the vast cathedral-like interior of this shed was used to hold fund-raising dances.

Gorinn

Gnarpurt

INGLEBY

INGLEBY, Winchelsea, Victoria

A designed shed that reflects the wealth of the grazier of that time.
Built in 1882 to replace an earlier shed, it shows a decorative use of bluestone that extended through homestead, stables, woolshed and shearers' quarters.
Although the use of such conscious decoration is unusual in woolsheds, the functional use of the shed was not overlooked, and the outside positioning of openings echoes the interior activities of the large T-shaped building.
Present owner: I. H. Steel.

Decorative gable ends terminate the three arms of this woolshed

Tottington

TOTTINGTON, St. Arnaud, Victoria

Tottington was taken up in 1844 by Laurence Rostron, son of a wealthy English cotton manufacturer. He had great vision and spent large amounts of money and energy in developing the property, importing Merino rams from Spain and America and stallions from England, and breeding good stock.

The first building was a little stone hut in which the shepherds lived while establishing the run. Later, substantial station buildings were added. Probably one of the earliest was the woolshed, built in 1846. The shed is built on a framework of unsawn timber felled on the site. The timber was jointed with wooden pegs and the low walls made of vertical red gum slabs, also split on the site. The roof was of bark, or it may have been thatched with reeds, but this was replaced with galvanized iron during the 1880s, which probably accounts for its survival when other similar buildings have fallen to the ravages of the weather. The woolshed remains basically the same today except for some changes demanded by modern wool handling technique and the A.W.U. It is still in use, and the shearers much prefer it to the more modern sheds which are often hot and draughty. Up to 28 blade shearers were employed and, under the subsequent ownership of Andrew Anderson, up to 55,000 sheep were shorn annually.

The large homestead, stables and other outbuildings were built of red brick baked on the property, with wooden shingles on the roofs. Workmen's cottages lined the river like a village but were often flooded, and they crumbled, leaving only the homestead and stone cottage which are on high ground, and the woolshed which, although on the flat, seems impervious to water.

Around 1852 Captain Melville, a gentleman bushranger, was in the area and his gang was believed to be stealing Tottington sheep for stores. Rostron, not to be thwarted, took every man on the station and rode off determined to bring in Melville, dead or alive. Only Mrs Rostron, her daughter and the domestics remained and, as night came on, a stranger rode up and requested a bed for the night. This was common practice at the time and travellers were put up in the men's quarters. But this gentleman was of such excellent address that Mrs Rostron invited him to stay at the homestead. Then followed a charming evening of music and conversation. The next morning the gentleman mounted his horse and, before departing, gave Mrs Rostron's daughter three little gold nuggets and asked Mrs Rostron to '. . . inform your husband that you have had the honour of entertaining Captain Melville'. One of the few occasions on which any man got the better of Laurence Rostron.

Present owner: Allan Small and family.

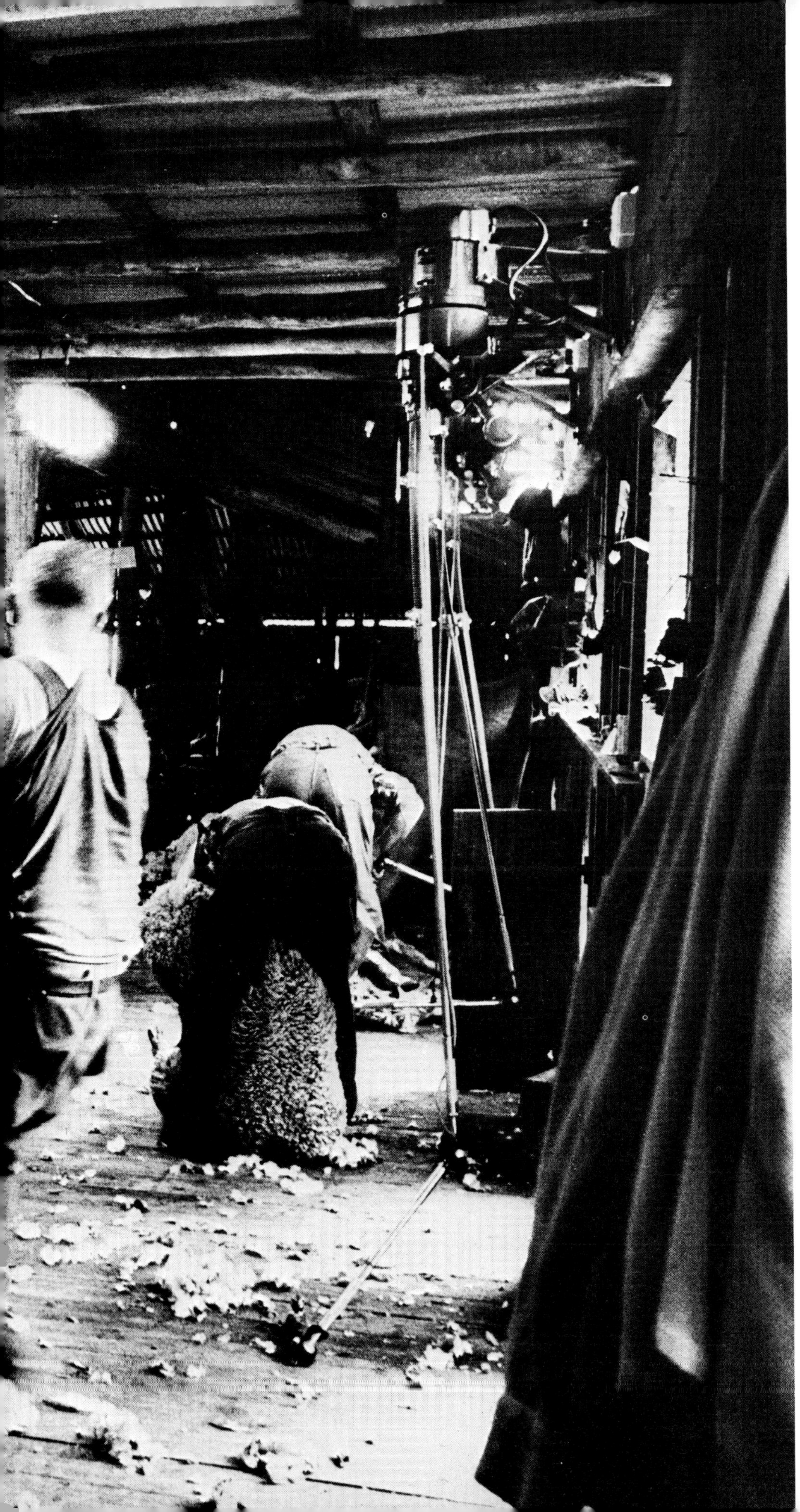

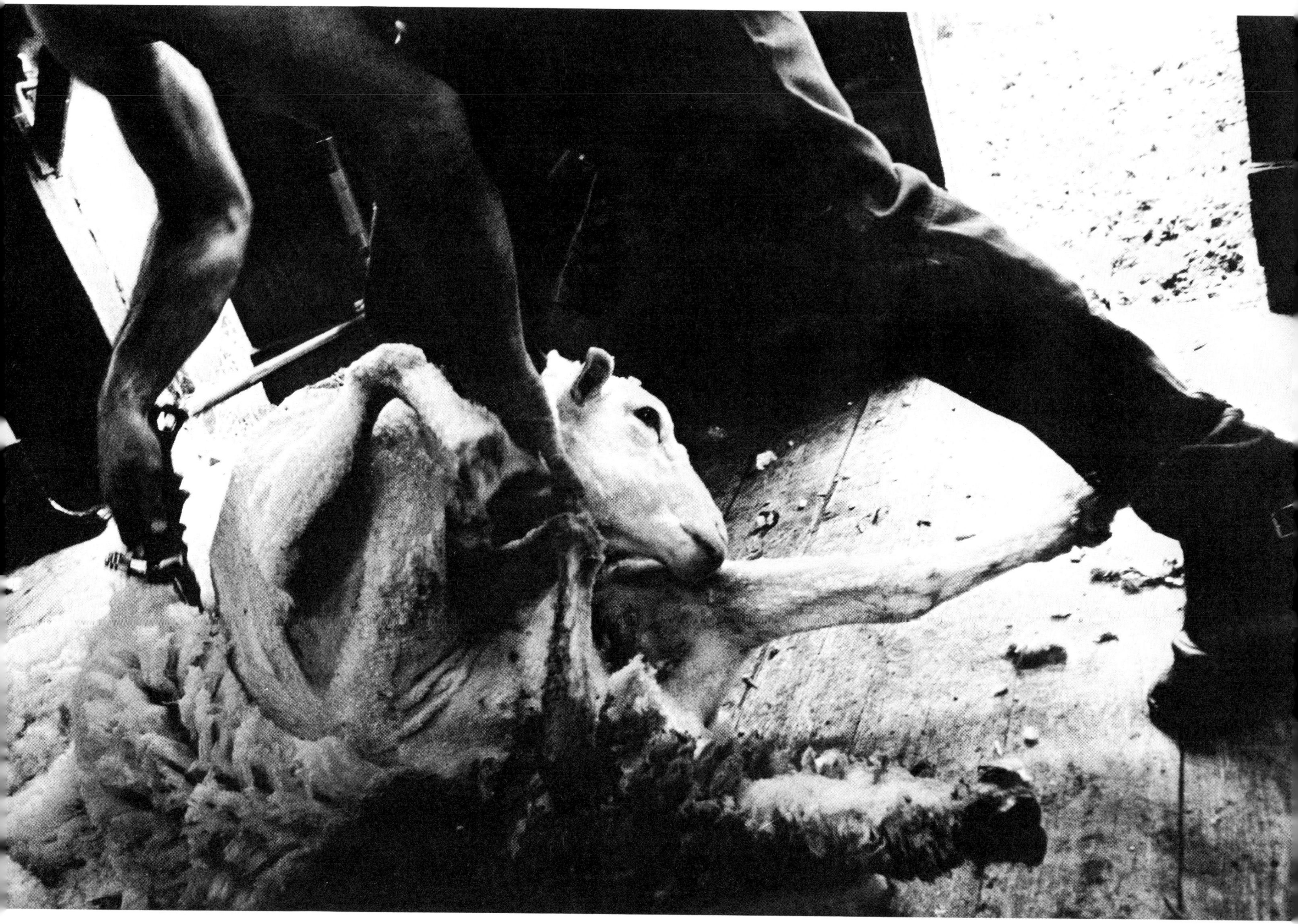

Bringing in the mob, Barwidgee

BARWIDGEE, Caramut, Victoria

Built of bluestone and bush timber, 225 ft. long, this shed stands on high ground alongside the creek that runs through the property.
The shed is thought to have been built in 1880, when 24 blade shearers were employed.
It now has 12 machine stands.
Present owner: Barwidgee Estate.

The Shearer's Workbench

It is here on the shearing board that the shearer bends to the monotonous, back-breaking task of removing the fleece from the sheep.

His length of stay in one shed is short and he takes each shed as he finds it.

Unlike the more permanent workplace of the city dweller, the shearer's does not become personalized.

He works within a space of five feet between shearers (in this shed a distance that contains catching pen and chute to counting pen) in which he catches, shears and releases the animal.

His few tools of trade as shown consist of
oil can,
nail clippers,
tin of oil containing brush for cleaning combs and cutters,
screwdriver for adjusting handpiece,
aspirins (bending all day is hard on hangovers),
spare combs and cutter
and the handpiece itself.

The markings on the timber strut, where combs have been smoothed, is the only evidence left of past shearers.

ASPRO SOLUBLE

BLYTHVALE, Streatham, Victoria

The old section of the woolshed (lower part on left of photograph) was built in the 1840s and extended in 1905 to a design based on an old shed at Polly Creek in the Scone area of New South Wales.
Box and stringybark timbers were brought from that area, by bullock wagon to Newcastle, shipped to Portland or Port Fairy, by rail to Beaufort, then by bullock wagon again to Blythvale.
Present owner: L. G. Weatherly.

Worn away by 130 years of sheep

MURDEDUKE, Winchelsea, Victoria

The shearers' huts stand like seaside bathing huts (this may have influenced the interior décor) as they face the bluestone and weatherboard shed.
This woolshed is of uncertain date but it is probable that the weatherboard part was moved from a nearby property in 1888.
Present owner: J. W. P. Wilson.

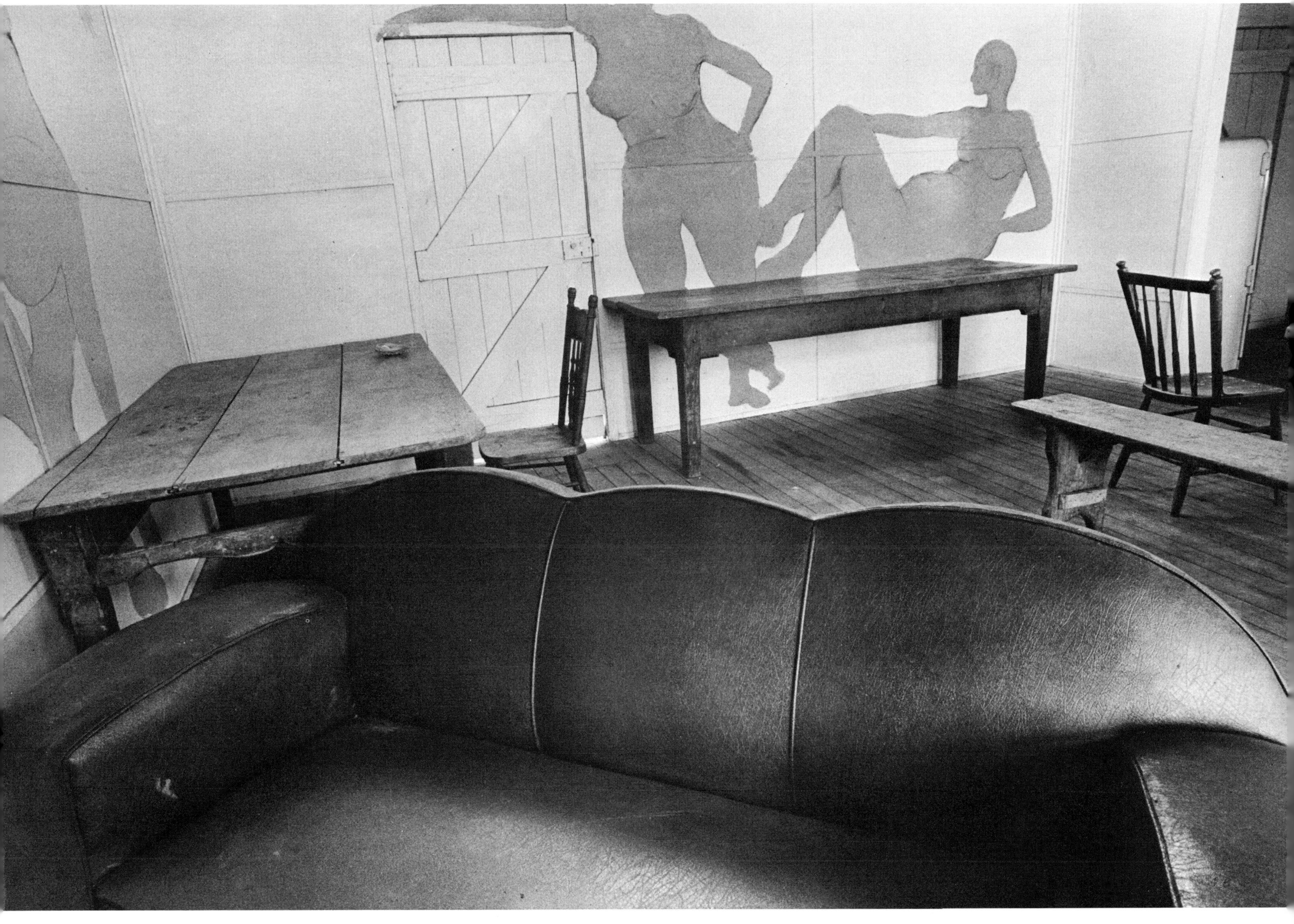

'There was three of them girls to every chap'.
from 'The Shearer's Dream', by Henry Lawson

BENAYEO, Binnum, South Australia

A 66 ft. x 22 ft. woolshed built of local ironstone by the original owner, Hugh McLeod, with the assistance of convict labour.
This was in 1850 and, at that time, the wool room was on an upper level above the arched entrance door.
One of the beams was put to melodramatic use in later years when the property manager hanged himself, his untimely death coinciding with the running of the Melbourne Cup horse race.
Present owner: D. L. & J. M. Cother.

A variety of stonework and richness of texture at . . .

. . . Hynam House, Hynam, South Australia

TARRAWATTA, Angaston, South Australia

Built in 1840 by the present owner's grandfather, the most interesting feature of this woolshed is the floor of the wool room.

Blue gum trunks measuring 3 ft. across have been shaped to a hexagon and driven some way into the ground, exposing the end grain of the timber. Years of contact with the grease from the fleeces have polished these logs to a rich sheen.

The building seems to be a shed within a shed, no doubt brought about by additions to meet needs for a larger woolshed.

Present owner: R. F. Angas.

PADTHAWAY, Naracoorte, South Australia

As at Tarrawatta, the wool room of this shed has an unusual timber floor.
Slabs of split red gum are laid end to end and secured by wooden pegs.
A space approximately 3 ft. square was left in these slabs and a hole dug to take the wool bale and, when the bale was full, it was raised by a winch and made ready for transporting.
These days the more conventional type of press is used in this 1850 stone building.
Present owner: A. K. Lawson.

Harry Pettit, an old blade shearer from Binnum, South Australia, said 'The oldest and largest woolshed I know is a big old red gum tree near here—bloke and his wife used to shear there for years—in the shade—she took the bellies off while he shore the rest.'

Bungaree

BUNGAREE, Clare, South Australia

The three brothers George, Charles and James Hawker settled what was at that time the most northerly settlement in the province, naming it Bungaree.
This was in 1841.
The property was improved, the homestead and other outbuildings built and the now famous Bungaree Merino Stud established.
Shearing sheds built during the first twenty years were unroofed, apart from a light brush shelter over the board to protect the shearers from the sun.
As the size of the flock increased, so the limestone woolshed grew to cope with the 44 blade shearers then needed.
A hundred extra hands would be employed at shearing time, and shearers' quarters had to be built to house up to 200 men, who greatly increased the amount of stores and provisions needed to support the establishment.
There are tales told of those times such as how, in the early 1900s, between five and six in the morning, a shearer cut his throat with his own blades—the blood-stained boards being replaced during breakfast.
Others remember the service that was held in the shed to bless the first fleece shorn, while the ceaseless labour of the shearers continued in the background.
When shearing was over and the shed cut out, the wool clip was carted by horse teams across country to Port Wakefield and shipped direct to England.
As in the case of many other sheds, the past is becoming forgotten, as a generation that only remembers machine shearing takes over properties.
Present owner: Sir Richard Hawker.

'In his cane bottom chair sits the boss of the board', from 'Click go the Shears'

A quality of light

A huddle of sheep, Queensland

A scattered spread, South Australia

DARR RIVER DOWNS, Longreach, Queensland

Standing alone in this vast, flat, sunburnt landscape, away from the noise and bustle of the showground where it originated.
The transformation from exhibition hall to woolshed was made when the building was brought from Melbourne to the Darr River property.
Reputed to have had 100 blade shearers, it is now equipped with 15 machine stands.
Record books show that in 1910 13,000 sheep were shorn at 5d. a head and woolpacks were 2/6½d. each.
Today's wage scale for shearing is $22.83 per 100 for stock sheep and $45.67 for rams.
Present owner: T. R. Lowry, New Zealand.

Oakwood

OAKWOOD, Charleville, Queensland

Believed to be the largest operating shed in Queensland, this woolshed can still utilize its 20 machine stands which are driven by the last steam engine in the pastoral industry.
The overhanging roof and sunshades of hessian protect the shearer from the blistering sun and induce an airflow through the shed, which is constructed of both machine-sawn timber and bush poles clad in galvanized iron.
Present owner: New Zealand & Australian Land Co.

Smoko

Tossing the fleece

Smudged activity on the board

BONUS DOWNS, Mitchell, Queensland

In 1962 the large Bonus Downs station was split up into five selections.
The woolshed, which is twelve miles away from the homestead, is now on the property of Alec Crick.
Although attempts have been made to give some shade to the shed from overhanging trees, the sun burns in from open verandahs and roof ventilators to light the vast timber interior.

Cut out

Isis Downs

ISIS DOWNS, Blackall, Queensland

Though not an indigenous shed (it was manufactured by Dorman Long, Middlesbrough, England) it is known as one of Queensland's 'greats'.
Prefabricated in steel and semi-circular in design, its web of steel trusses stretches out to the far corners of the shed.
Built in 1908 for 52 machine stands, with the catching pens around the perimeter of the half-circle, each shearer obtained full benefit of uninterrupted clerestory light.
Powered by electricity from the start, it boasts two hydraulic wool presses.
Present owner: Clarke & Whiting.

Acacia Downs

ACACIA DOWNS, Aramac, Queensland

The fact that not all woolsheds start life as such is certainly true of the one at Acacia Downs. This shed began as a meat works at Hughenden in 1898, but it was to fulfil this function for one season only, as a prolonged drought of four years put it out of business. Part of the meat works, containing an overhead water-tank, was dismantled and brought down by horse team to the property sometime during the 1920s.
In the shed's present role shearing is carried out on the upper level, sheep being brought up on a slate ramp and, after being shorn, deposited down a chute.
The wool clip awaits transport on the ground floor and the press links the two levels.
Present owner: The McLeish family.

Looking down on the board and related areas

NORTHAMPTON DOWNS, Blackall, Queensland

Two drovers, their cook and 14 dogs had stopped overnight at Northampton Downs shed.
This was the first drove for two years because of drought conditions and they were moving 4,500 sheep 160 miles at the rate of four miles a day.
Fitted with 58 machine stands in 1889, this shed became the first big Queensland woolshed to be converted from blade shearing.

Yarralumla

'At Canberra, the Yarralumla woolshed, wholly built of galvanized iron, is still one of the more successful buildings, in my view, in the National Capital.
Certainly, I could willingly see several dozen present-day buildings in Canberra demolished so long as that woolshed was preserved.'
His Excellency The Governor-General
The Right Honourable Sir Paul Hasluck, G.C.M.G., G.C.V.O., K.St.J. speaking at the Centenary Convention of the Royal Australian Institute of Architects, 1971.

The prize fleece . . .

. . . Annual Sheep Show, Sydney

Sale yards, Blayney, New South Wales

Following close in the tradition of woolshed architecture are the other allied buildings such as sale yards and stockyards.

Stockyards, Belltrees, Scone, New South Wales

- 18 59 -
WOOLSHED - INN
BAR
PENFOLDS
PORTS
SONS

WOOLSHED ST
POST OFFICE

FERMENTED
SPIRITOUS
Tooth's
'SHEAF'
STOUT

'All on the board'. Called when the last sheep of the mob are in the catching pens.
Bare-belly. A sheep with all wool rubbed or dropped off its belly.
Board. That part of the woolshed where the shearer works.
Boggi (pronounced bog-eye). A shearer's handpiece shaped like the Boggi lizard.
Boss's boots. Because his vision is restricted when bent over shearing, this is what a shearer watches out for.
Bowyangs. Straps or string around a shearer's trousers, under his knees.
Broomie. The broom hand on the board.
Catching pens. The pens from which shearers catch their sheep to shear them.
Chinaman. A lock of wool left unshorn on the sheep's rump, and not noticed by the shearer until the sheep is disappearing through the porthole.
Classer. The wool expert, who grades the fleece before it is pressed.
Cobbler. A sheep difficult to shear, and usually left to last.
Cut out. The end of one mob of sheep, or the end of shearing.
Deucer. A shearer who regularly shears 200 sheep per day.
Drummer. The slowest shearer on the board, who shears at the bottom end of the board, farthest from the wool tables.
Expert. The man who grinds the gear and keeps the handpieces and plant in order.
Flyer. A fast sheep to shear, or the best in the mob.
Gasper. A sheep gasping for breath while being shorn, usually because of grass coming up into its throat.
Gong. Usually a piece of metal hanging by the wool table, struck to notify start and finish of a run.
Gun shearer. One who has reached the top, with outstanding ability.
Kicker. A sheep that keeps on kicking while being shorn.
Moccasin. Shearer's footwear, originally made from wool bags but now usually of leather.
Picker-up. The man who picks up fleeces and takes them to the wool table.
Porthole. The opening through which sheep are released after being shorn.
Ringer. The fastest shearer on the board, usually shearing on the first stand next to the wool table. Any rival shearer must shear a bigger tally for three consecutive days before he can take the ringer's stand.
Rouseabout. The shed's general hand who has a great variety of jobs.
Run. The shearing time worked between official stops, smokoes or meals.
Sandback. A sheep with sand, grit or dirt in the back wool.
'Sheepo!' Shearers call 'sheepo' when they have caught the last sheep in their pen, to let the rouseabout know the pen is empty.
Shiack. The shed banter, in which all take part, from boss to rouseabout. The good-natured teasing does much to make a hard job a lot easier.
Smoko. The much looked-for morning and afternoon tea.
Snowed in. The state of affairs when shearers get ahead of the wool table and wool lies around waiting to go on the table. On a good run shearers sometimes snow the pressers in and this leads to their shearing faster and faster, as they enjoy getting ahead of the rest of the shed.
Sweat rag. A towel used to wipe sweat off during a run.
Sweating pens. Pens in which sheep are left to cool after mustering and before shearing.
Tally. The number of sheep a shearer shears in a day.
Tar boy. The hand who puts a smear of tar on cuts made on sheep.
Tassel. Greasy lock of wool left under the legs and brisket.
Two cut. A second cut required when a staple of wool is cut above its base and needs trimming. Such short pieces are of little value.
Up in the air. Said when a shearer flashes his handpiece around in the air, cutting more air than wool.
'Wool away!' The shearer's call to the picker-up when a fleece has not been cleared from the board and is in the shearer's way.

ANDERSON, RONALD. *On the Sheep's Back. Past, Present—Future?* Rigby, Melbourne, 1967.

ANGUS, J. COLIN, and FOSTER, HARLEY W. *The Ovens Valley*, Cypress Press, Melbourne, 1970.

AUSTIN, H. B. *The Merino, Past, Present and Probable* (3rd edition). Grahame Book Co, Sydney, 1944.

BEATTY, BILL. 'A Treasury of Australian Folk Tales and Traditions.' *Walkabout*, Sydney, 1968.

BISCHOFF, JAMES. *A Comprehensive History of the Woollen and Worsted Manufactures & The Natural and Commercial History of Sheep, from Earliest Records to the Present Period.* Smith, Elder & Co., London, 1842.

BRODRIBB, W. *Recollections of an Australian Squatter.* John Woods, Sydney, 1883.

BULL, W. *Jondaryan NSW—History.* Griffith Historical Society, 1960.

BURFITT, C. T. *History of the Founding of the Wool Industry of Australia* (2nd edition), F. W. White, Sydney, 1907.

CALVERT, 'View of Sheep Shearing.' *Illustrated Sydney News*, 15 October 1864.

CAMPBELL, W. S. 'Historical Sketch: The Establishment of the Sheep & Wool Industry in Australia.' *The Sydney Stock and Station Journal*, 25 June 1912.

CARTER, H. *His Majesty's Spanish Flock: Sir Joseph Banks & The Merinos of George III of England.* Angus & Robertson, Sydney, 1964.

COLWELL, REV. J. *The Story of Australia*, Vol. 4. Clarke Publishing Co. Australasia Ltd, 1925.

COX, E. W. *The Evolution of the Australian Merino.* Angus & Robertson, Sydney, 1936.

DOW, HUME (Ed.). *Trollope's Australia.* Nelson, Melbourne, 1966.

FISON, E. H. *Flocks and Fleeces: Being a Concise History of the Sheep and its Wool in All Countries.* Wyman & Sons, London, 1894.

FOSTER, HARLEY W. *The Central Gold-fields.* Cypress Press, Melbourne, 1969.

FRASER, A. D. *This Century of Ours.* Halstead Press, Sydney, 1938.

FREELAND, J., COX, P. & STACEY, W. *Rude Timber Buildings in Australia.* Thames & Hudson, London, 1969.

GUNN, D. & NILLS, J. *Links With the Past.* Brisbane, 1937.

GUNN, J. 'The Terminology of the Shearing Industry.' *University of Sydney Australian Language Research Centre, Occasional Paper No. 5.* Sydney, 1965.

GUTHRIE, J. F. *A World History of Sheep and Wool.* McCarron Bird, Melbourne, 1957.

HADDON RIG. 'Early History of the Famous Stud.' *New Nation Magazine*, June 1926.

HARRIS, W. K. *Outback in Australia.* Garden City Press, Letchworth, 1913.

HODGSON, C. P. *Reminiscences of Australia With Hints on the Squatter's Life.* W. N. Wright, London, 1846.

JARVIS, J. *Condobolin: The Story of its Development (Burrawang Station).* Cumberland Newspapers Ltd. Parramatta, 1940.

'The Jondaryan Woolsorter.' *The Queenslander.* Brisbane Courier General Printing Office, 1868.

KEATINGE, COL. MAURICE. *Travels Through France and Spain to Morocco.* Henry Colborn British Foreign & Public Library, London, 1817.

LAWSON, R. A. S. *Umbiwarra. Reminiscences of My Life on Stations in NSW and Queensland Seventy and Sixty Years Ago.* R. A. S. Lawson, 1937.

'The Birth of the Australian Wool Industry: Macarthur and His Achievements.' *P.F.A. Quarterly Magazine*, June 1814.

MCIVOR, C. *History and Development of Sheep Farming From Antiquity to Modern Times.* Tilghman and Barnett, Sydney, 1893.

Mackaness, G. *Sir Joseph Banks (Banks, Macarthur and the Wool Industry).* Angus & Robertson, Sydney, 1936.
Mathews, W. J. O. H. 'Construction of Shearing Sheds and Yards.' *NSW Agricultural Department Farmers' Bulletin 91*. 1915.
Mathews, W. J. O. H. 'Shearing Sheds for Small Flocks.' *Agricultural Gazette. NSW*. January 1915.
Mills, A. & R. *Sheep-O! The Story of the World's Fastest Shearers*. Reed, Wellington, 1960.
Munz, Hirsch. *The Australian Wool Industry*. Angus & Robertson, Sydney, 1950.
Niland, D. 'Champion Shearers of Australia.' *Walkabout*, February 1943.
Ogilvie, W. H. *My Life in the Open* (Colonial Ed.). T. Fisher Unwin, London, 1908.
Roberts, S. H. *The Squatting Age in Australia 1835-47*. Melbourne University Press, 1935.
Sharp, K. & Crump, B. *A History of Milo and Ambathala*. Milo Pastoral Co. Ltd., 1963.
'Shearers' Strike, Queensland, 1891.' *United Pastoralists Association of Queensland*. Muir & Morcom, Brisbane, 1891.
'The Shearing Industry: a Century and a Half of Development.' *Country Life Journal*, Sydney, 1 October 1937.
'Sheep Shearing in Australia.' *Illustrated Sydney News*, 28 January 1854 and 16 October 1864.
'Sheep Shearing at Yanco.' *Illustrated Sydney News*, February 1869.
Sorenson, E. S. 'In the Days of the Blades.' *The Land, Farm and Station Annual*, 1939.
'South Australian Merino Flocks: Evolution of Breed in Hand of Stud Masters.' *Centenary Review*, 1936.
Stuart, Julian. *Shearers' Strike, Queensland, 1891*. Australasian Book Society, 1967.